THE HEALING POWER OF
SMUDGING

Cleansing Rituals to
Purify Your Home,
Attract Positive Energy,
and Bring Peace
into Your Life

Rodika Tchi

Ulysses Press

To the power of love in you.

Published in the U.S. by:
Ulysses Press
P.O. Box 3440
Berkeley, CA 94703
www.ulyssespress.com

ISBN: 978-1-61243-760-6
Library of Congress Control Number: 2017952131

Printed in the United States by Bang Printing
10 9 8 7 6 5 4 3 2 1

Acquisitions editor: Bridget Thoreson
Managing editor: Claire Chun
Project editor: Shayna Keyles
Editor: Renee Rutledge
Cover design: Chris Cote
Interior design: what!design @ whatweb.com
Photos: © Rodika Tchi except burning sage on cover
    and pages 1, 5, 15, 24, 47, 54, 72, 73, 74, 82 © Sophia Vartanian
Layout: Jake Flaherty

Distributed by Publishers Group West

CONTENTS

INTRODUCTION

When you hear the phrase "smudging your home," what images come to mind? What thoughts and feelings? Do you feel a slight sense of intimidation? Does it sound too complicated? Messy? Unnecessary?

Even though the ritual of smudging has made a comeback in recent years, many people remain daunted by the concept of smudging their homes. The idea can leave one feeling skeptical or overwhelmed, either because they fear creating too much smoke, believe that smudging is somehow a religious ritual, or feel that the whole process is too complex.

Even when one has the intent or the enthusiasm to smudge, it is easy to abandon the process of smudging at the initial stage of buying a smudge stick! I have seen many forgotten smudge sticks in clients' homes because they felt that smudging was too complicated to even try.

As with anything new, it is helpful to know the benefits of any process you are planning to do. Why should you smudge your home? Do you actually need to smudge it? Can you do this in a way that is simple and enjoyable? Are there any other ways to clear the energy in your home? And what makes smudging so special?

Just like in dating, the more questions you ask, the better—and sooner—you will know if it is worth investing in this relationship. I use the word "relationship" because, ideally, entering the world of smudging means entering a beautiful and long-term relationship. As with any relationship, you go through stages in order to get to a place of deep intimacy and commitment.

In this book, I will help you answer these questions and share many different ways to smudge, including some really simple techniques that allow you to smudge even several times a day with much enjoyment. I have been smudging my home for many years; sometimes I smudge my home simply because the scent of sage makes me feel centered and peaceful. I have grown to truly love and admire the spirit of this herb, and experience its energy as an intimate, loyal, and cherished presence.

As I am a feng shui consultant, I can immediately feel how the energy in my home can shift after I have been out all day and bring home busy, hectic, and sometimes negative energies from outside. In this situation, a quick two- to three-minute smudging can do wonders to calm and purify both the space and my own energy.

The beauty of smudging is that it can be as simple and quick or as complex and elaborate as your heart desires. Once you know the basics and all the optional add-ons, you will feel more empowered to create your own rituals to fit any situation. The process of smudging can be simple, enjoyable, and deeply centering and grounding. The main benefit of smudging, however, lies in its ability to purify the energy of your home in order to help you live with vibrant, happy, and peaceful energy.

I promise you that by the time you finish reading this book, you will feel confident, empowered, and ready to take charge of the energy in your home, and in your life, by smudging. You will feel excited and enthusiastic about creating a whole new level of good energy in your own home. As a bonus, you might also learn some valuable tips about creating good feng shui in your space.

Chapter 1

WHAT MAKES SMUDGING SO POWERFUL?

Let us start by clearly defining smudging. In a nutshell, smudging is the process of burning various sacred substances such as herbs, resins, or woods in order to purify the energy in any given space. Smudging is also used to purify one's own energy, as well as items such as jewelry, decor items, and even clothing. This purification occurs when the smoke produced from burning these sacred substances spreads into the air.

Recognizing the quality of specific energies in your home in order to know when to smudge is not difficult. All energies express themselves in specific feelings. Do you start feeling sad as soon as you enter a space? Maybe tired, or even upset? Are there rooms in your home that you avoid spending time in?

When it comes to understanding energy, your body always knows how to read it. It does so very quickly and it never lies, so it is good to listen to your body's feedback as soon as you enter a space. What you feel in the first minute or so upon entering any space will tell you a lot about the presence of specific energies and their core signatures.

The power of fire helps transform your life

There are many techniques for purifying your space, most commonly known as space-clearing techniques. These techniques use movement, various vibrations of sound, and the power of water, words, and scents. The most powerful space-clearing rituals always involve the power of fire. Even in the form of a small burning candle, fire brings what no other element can—swift purification and illumination. Sitting around the fire was a sacred ritual in many ancient tribes and communities, and the benefits provided by fire go far beyond physical warmth. Fire illuminates and inspires, it warms our hearts and purifies our energies. It connects us to the energy of Spirit, it brings trust and meaning to the often chaotic human life.

Fire dispels not only visual darkness, but more importantly, energetic darkness. It clears both your personal energy field and the surrounding space of stagnant energy, painful memory residues, negative attachments, and a vast array of energetic debris in your aura. There are many purifying and calming meditative techniques in which one is encouraged to look into the flame in order to move deeper and deeper to a place of power, trust, and strength.

Even though all elements are considered sacred, the fire element, represented by the sun, possesses a unique, life-giving power. Language referring to the soul has always included

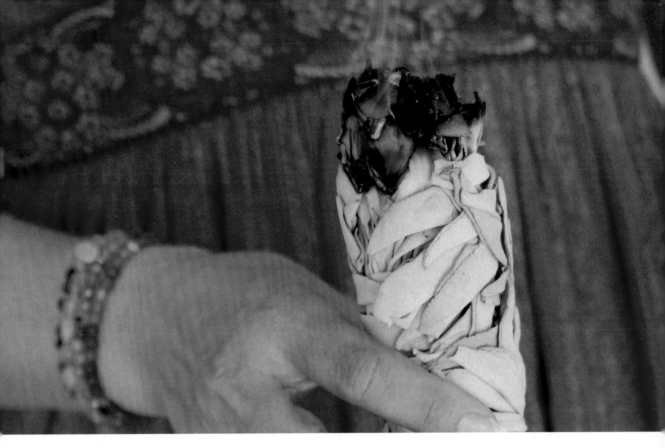

fiery terms—the flame of the soul, the spark of life, etc. Fire is also the only element that moves in only one direction—upward toward the sky—and thus it is considered to have a direct connection to the Spirit, Source, or the Ultimate Creator.

When living in Bali, one of my most cherished memories was walking with a friend late one afternoon and suddenly smelling the scent of incense coming from every direction. We were on a road in an open field, the sun was setting gently, and the soft wind was bringing that aroma in waves so as to softly caress us with it. Even though it felt like I was transported to a magical place, I was also very present to everything around me. The exhaustion of the long journey to Bali lifted off my shoulders as if it had never existed.

Such is the power of fire; it can create truly transformative energy when skillfully combined with herbs, resins, or woods. Of course, fire can also burn and destroy; in its extreme expression, it becomes a force of wrath and destruction. Knowing how to handle the spirit of fire requires respect, patience, and deep humility. This is the main reason why only shamans and elders performed purifying ceremonies in ancient cultures. They had the inner power to handle fire.

Smudging has ancient roots

Because of the unique power of fire, cultures around the world have used various fire rituals—some very similar to smudging—for a very long time. We tend to think of smudging as a Native American ritual, which is certainly true, but the ritual has been practiced throughout the world, from Bali to Hawaii, from New Zealand to Ireland. These smudging rituals used a wide variety of substances, depending on what was growing in the area. Frankincense and myrrh were abundant in Egypt, while Celtic people used herbs such as vervain, mugwort, and juniper.

Native American smudging traditions

The Native American tradition of smudging varies from tribe to tribe, but essentially, each smudging ceremony has a similar structure. Different tribes use different herbs, depending on their availability. Although sage is the most well-known smudging herb today, not all Native American tribes use sage. Other popular herbs are sweetgrass, cedar, and tobacco.

Traditionally, the herbs are placed in a bowl and burned. A hand, feather, or smudging fan is used to fan the smoke of these burning herbs. The tradition of bundling various herbs together in smudge sticks or long braids is also widespread because of its convenience.

A smudging ceremony is always performed with humility and reverence. It is a time for prayers and blessings, for purification and connection to the Great Spirit. This ceremony could be connected to a bigger event, such as the Sun Dance or a vision quest, or it could be a shorter ritual completed before healing a patient.

Beautiful prayers asking the Spirit to bless human life are always part of a smudging ceremony, from asking to open one's heart in order to deeply feel the Creator's love to asking for the ability to walk in truth, strength, and courage. Specific prayers are added by the elders based on the needs of the community.

All parts of the Native American smudging rituals—from asking for the protection of spirit guides to reciting beautiful prayers—have powers in them that you can apply in your own smudging rituals.

The power of cardinal directions and spirit animals

Most ancient healing and smudging rituals begin by honoring the power of the cardinal directions. This act acknowledges our connection to Mother Earth and the cosmos at large. Honoring the four cardinal directions evokes humility and awe of the powerful universal forces at play that constantly influence human life. In the ancient art and science of feng shui, the four cardinal directions, the directions in between, as well as the center of the space are taken into consideration.

The energy of the cardinal directions is expressed in a multitude of ways, such as in specific elements (water, fire, air, etc.), seasons, plants, colors, and even sounds. It is important to understand that these associations are culturally specific, rather than universal, so it is wise to take some time to find your own way of honoring these powerful energies. If you try to follow one set way, then read about another way of honoring the directions, you might feel confused. For example, in some cultures, the direction of East is represented by the color yellow because it is the direction of the rising sun. In the ancient art of feng shui and geomancy, though, the East is represented by the color green, which expresses the sun's beneficence in allowing all life on earth to flourish. Are either of these representations better than the other? Of course not. There are so many ways to express any given energy, so it is not a matter of right or wrong, better or worse.

What gives power to any ritual, including a smudging ritual, is the personal expression of a conscious choice. To bring sacredness and genuine power to your ritual, find a way to clearly and authentically express a specific energy in the manner that is closest to your heart. As I am a feng shui consultant, I naturally tend to connect to the very essence of any specific energy and then find its expression in nature's elements. I love to dive deep to reach the core signature of any specific energy, and then find the most accurate feeling that connects me with it.

When I evoke the power of the **East** direction, I honor the joy of the rising sun and express gratitude for the blessings it brings to all kingdoms, from mineral to human. I feel the childlike innocence in my heart before the creation of a new day. I see the morning dew drops softly melting on the fresh green grass, flowers opening to face the sun, and tree leaves dancing with glee in the shimmering morning sunlight. I rejoice in this feeling and I ask for help in consistently cultivating this joy in my heart.

South is the direction of the strong fire element in feng shui. It is the majesty of the sun at its highest peak, the expression of full illumination and unquestionable power. When I evoke the help of the South direction, I surrender to this tremendous power and ask for help in illuminating and cultivating my own power. I ask for the courage to shine bright, to trust the voice of Spirit, and to be grounded in my own inner power.

West is the direction that holds the wisdom of a well-lived life. It holds, with crisp clarity and soft wisdom, the lessons of a long journey. I connect to a warm energy that is benevolent and kind, and I ask for its guidance. Would I regret not taking specific paths at the end of my life? What is the best way to proceed in a specific situation, project, or relationship I am in now? West is the direction that already has all the answers for you; it is the future version of yourself reaching out and helping with guidance, if you are ready to ask and listen.

And finally, **North** is the direction of deep inner reflection. It is the wisdom of the water element, water that has experienced all of its stages and is now coming back to itself for a time of reflection and silent regeneration. In feng shui, this direction holds the energy called "path in life." When I connect to the direction of North, I bow and become very still. I open to the deep, silent wisdom of this incredibly powerful element and try to sense what I can learn from it that is relevant to my own path at this moment in time. I invite its wisdom and versatility, and ask for the most appropriate expression to guide my heart.

My favorite way of honoring the sacred directions comes from a beautiful Ojibwe, or Chippewa, tradition where not four, but seven directions, are honored and invoked during a smudging ritual. We start by honoring the East direction, then proceed to the South, the West, and the North directions, and then look up and open the arms to honor celestial energies by honoring the Father Sky. We honor Mother Earth for all her love and bounty by gently touching her, and then we weave all these energies into our heart, which we express by living in power and truth. These three directions—Father Sky, Mother Earth, and Human Heart—are the additional directions that are honored along with the cardinal directions.

In many cultures, including the cultures of various Native American tribes, the energy of a specific direction is expressed by a power animal, also called a spirit animal. The energy of a power animal serves as a wise guardian, or gatekeeper, of a specific direction. As with colors and elements, there are many ways to express the energy of a specific direction, and the best way is to find your own power animals to guide you. You can also see if the traditional expressions listed below hold energy for you, and if you strongly resonate with a specific power animal energy in a specific direction. Here are the power animals that correspond with the four cardinal directions; as you can see, there is more than one spirit animal for each direction.

East—Eagle, Hawk, Condor

South—Coyote, Wolf, Rabbit

West—Bear, Crow

North—White Buffalo, Owl

If you want to work with the spirit of power animals in your smudging ceremonies, take some time to feel which one holds the most energy for you for a specific direction. Which spirit animal would you call to for its presence and helpful energy? You can explore the symbolism of specific animals to see which ones you feel drawn to the most. You can also work with your own power animal during your smudging rituals and see how it all unfolds from there. There is no rush with this process. You might already have the answer as you are reading this, or you might receive guidance months from now.

As always, choosing to trust in your inner voice above all others is the best way to go. Take the information that resonates with you and leave the rest; if you are meant to work with any specific energy, it will come back to you at the right time.

Smudging brings clearing, blessings, and protection

The process of smudging your home, especially when done regularly, will create calm and clear energy in your home. No matter which form of smudging you choose from the options in this book—from herbs to woods to resins—you are working with powerful elements from nature. These natural substances offer their power to transmute all the negative and stagnant energy that is often present in our homes. Along with clearing low energy, they have the power to bless and protect. Mother Nature, beautifully complex, is always there to give us more, to reveal more secrets, to open our eyes to simple ways to live happier lives.

The energy in your home affects your health and happiness

We have all heard many times by now that everything is energy. You, your house, your car, the neighbor that annoys you, your favorite flowers—absolutely everything is energy. We are all living in an intricately woven collective web of energy that constantly affects us, and we affect it with our thoughts and emotions. One of the most important factors in this dance, the factor that can be either an ally or an enemy, is the energy in your home.

Your personal energy affects all areas of your life, including your ability to manifest your dreams. The energy in your home has to be clear, strong, and nourishing in order to support you in all your life's endeavors. Your health and happiness depend on the quality of energy in your home.

Regular smudging rituals help clear the energy and provide a good energetic foundation for a happy home. Smudging also brings deeply nourishing energy into a home, a quality of energy that is both soothing and reassuring. This energy has deeply healing effects on all who live there, because it brings powerful vibrations from nature, an energetic kind of food that our bodies long for and need in order to keep healthy in this world.

Ultimately, we all long for a peaceful, calm place in which to relax and get in touch with ourselves, as well as the ones we love. The only way to reach that state is to realign our energies with the rhythms and power of nature. Smudging is a beautiful and easy way to bring these rhythms into our homes on a regular basis. Regular smudging rituals realign our energy, heal it, and strengthen the necessary energetic protection around our bodies and our homes.

Cautions before you decide to smudge

Many people feel uncomfortable with the idea of smudging their homes because the process of smudging involves smoke. Is this a good reason to avoid smudging, though? No, definitely not, but you should make an informed decision based on your own circumstances. In order to make a fully informed decision, you need to know both the pros and the cons of smudging.

Here are the three main cases in which it is best not to smudge.

1. If you or someone in your family is allergic to sage or any other herbs you plan to use, it is best to avoid smudging. It goes without saying that you should be extra cautious with smudging if you or anyone in your family has asthma or respiratory problems. It is also best to avoid smudging when you or someone in your family is pregnant. You can smudge when you know they will be away for a couple days, but do not do that without their permission if you share common space. Little babies and older people might not enjoy the scent of smudging, but this varies. In all these cases, you can try some alternatives to smudging, which I cover in detail in Chapter 7.

2. It is best not to smudge in a space where you cannot open the window so that the lower energy can be released. If the space has no windows or external ventilation, look into alternatives to clear the energy. When you are smudging a space with closed-circuit air conditioning and no windows, the benefits of smudging will be dubious, as the lower energy will just keep recirculating. In addition, the scent of smoke might, in time, create an unpleasant odor because it cannot be cleansed by the incoming fresh air.

3. It is not recommended to smudge someone else's space or energy without their permission. You cannot give a smudging ritual as a gift unless the person clearly asks for it. Please keep this in mind if you plan to apply all you will learn in this book to become an avid smudger of other people's places! Even with the kindest of all intentions and the most pure heart, you might be infringing on someone else's life and lessons.

I remember a friend telling the story of visiting her parents after a long time of being away. One evening when her parents were out, she thought to clear the energy in the living room, as it felt very "sticky." Even though her intent was definitely pure and beautiful, it turned into a disaster. As soon as her parents came home and settled into their usual evening routine in the living room, their interaction suddenly exploded into the worst fight they had ever had. Once you start working with energy, you will clearly understand why this event happened. Whatever energetic "stickiness" was in the space, it was her parents' to have and experience, so a drastic, uninvited clearing of it caused a lot of pain.

Please trust me and approach both the power of smudging, as well as everyone's right to their own space and own lessons, very seriously. Ask for permission if you want to help someone with clearing their space, and respect other people's right to say no. You might not know what is best for others at any point in time of their journey. Being able to shift the energy is both a gift and a responsibility. Please do not take it lightly.

Frequently asked questions about smudging

Here are answers to some of the most popular questions about smudging. This section should help put your mind at ease if you feel nervous or uneasy about smudging your space.

How long does it take?

If you have never smudged before, the initial smudging might take anywhere from 10 to 25 minutes, depending on the size of the space. Once you begin, the basic smudging sessions tend to become shorter and easier. My daily smudging does not take more than 5 minutes.

What time of the day is best for smudging?

Any time is a good time! I do some light self-smudging in the morning, when I am up early and feel like smudging with sage or palo santo to start the day. I also do some smudging at night if I had a busy day visiting many places. If you are a beginner or if you need to smudge a place thoroughly because of an illness, argument, or for any other reasons, it is best to use a window between 11 a.m. and 1 p.m., when the energy is most potent. (See more on the best times for smudging on page 32.)

Do I need to do anything specific before smudging?

Centering yourself and gathering your energy is very helpful before you start smudging your space. This process can be as simple as taking several deep breaths and saying a prayer, or something more elaborate like a long yoga session followed by a meditation. Your intent is what matters the most.

Where exactly do I start?

If you have an altar with set energies—meaning an altar that has been there for a while—it is always best to start at your altar. Otherwise, just find a convenient space for your supplies, ideally close to the center of your home.

Can I play music?

You can play music if you have something specific in mind that will help the process. Examples of music that can work are Native American flute music or soulful songs. More often than not, though, it is best to smudge in silence so you can connect and listen to your home. If you are carried away by the music, you might miss important messages from your space. Think of the time you spend smudging like the time you spend connecting with your beloved—you need to be in silence in order to truly hear the heart of another. But to begin with, use music if the idea appeals to you. Stay curious and attentive to see how each room in your house feels and reacts differently to this process.

Is there anything specific I need to do after smudging?

If your place felt heavy and you released a lot of energy during smudging, it is helpful to do one (or all!) of the following:

- Spend some time by yourself after smudging. Spend at least 5 to 15 minutes alone, and do not rush back into activity. The way you handle your energy right after smudging affects the way the energy will settle into your space. All three energies—yours, your home's, and the combined power of fire and herbs—are still deeply intertwined after smudging, so give them some time to settle and dissolve the energetic threads.

- If you received any insights during the process of smudging, consider journaling, or simply sitting in silence in front of your altar. If you have done a major smudging or clearing session, you can take a shower or wash your hands and arms up to your elbows in cold water.

🪷 Go for a walk in nature, ideally to a place surrounded by old trees, in order to help you ground and circulate your energy.

Chapter 2

HOW SMUDGING CAN HELP YOU

Smudging helps clear your energy and the energy in your home, as well as bring the blessings you might be longing for in your life. You can use smudging rituals for specific occasions, such as moving into a new home or clearing the space after heavy arguments. You can also use daily smudging rituals to bring more clarity and peace into your life. Let us have a closer look at the benefits of smudging.

Smudging purifies the air and allows a fresh start

Several studies have shown that smudging kills airborne bacteria and acts as a natural air purifier. The *Journal of Ethnopharmacology* published a study that found that medicinal smoke can eliminate bacteria by 94% in 60 minutes.[1] A mixture of woods and medicinal herbs were used in this study, and the amazing discovery was that along with drastically reducing the presence of airborne bacteria in as little as an hour, seven of the pathogenic bacteria were still not detected in the room almost a month later! This research concluded that by using medicinal smoke, it is possible to eliminate numerous pathogens from the air of any confined space.

As a feng shui consultant, I have rarely seen a home that does not have energetic blockages and pockets of stagnant energy. This situation is the result of many factors ranging from constant expression of negative emotions to a poorly designed floor plan. The main cause, though, is our ignorance of the way the energy in our environments works, and how important it is to clear energies on a regular basis. If we do not clear our homes on an energy level—just like we clear our homes on a physical level—there will always be unpleasant and stagnant residues that promote a sense of unhappiness at home.

By using the power of natural elements, smudging can remove energy blockages in your space by either dissolving or transmuting them, thus allowing for a fresh flow of energy. And you know what a fresh flow in your home does—it lends you the energy to create a fresh, new start in your life.

---

1   C. S. Nautiyal et al., "Medicinal smoke reduces airborne bacteria," *Journal of Ethnopharmacology* 3, no. 144 (Dec 2007): 446–51, doi: 10.1016/j.jep.2007.08.038.

The feng shui of your home needs smudging

Feng shui is a fascinating subject. The more I practice it, the more amazed I am at the simplicity of its power. I know it might sound contradictory to use this word in relation to feng shui, but trust me, the basis of feng shui is simple. If we had the ability to chat with the ancient feng shui masters, I know they would say something very simple. It might be as simple as encouraging us to learn from nature and to notice how we feel when outdoors as compared to the feeling we have when we spend time at home. Once you are consciously aware of the difference between nature and the man-made world, feng shui encourages and gives you the tools to replicate the vibrant feeling of nature in one's own home.

The whole premise of feng shui is to bring the wisdom, vibrancy, and healing power of nature into our homes. We have forgotten this simplicity somewhere along the way and have instead become lost in feng shui calculations, exotic images, and quick fixes. The truth is that if the energy in your home is stale and stagnant, no amount of feng shui cures can help. Of course, there are many aspects to what makes a good feng shui home, and while smudging is not a panacea for all woes and will not create good feng shui in an instant, it will give the energy of your home a very healthy boost. And if you smudge regularly, you will create a good and healthy feng shui foundation for your space.

Smudging and the feng shui of specific rooms

Feng shui has been called the acupuncture of a house for a good reason. Each room in your house serves a specific energetic function, and each room has a different ability to deal with the negative energy expressed or stored there. Some areas of your home allow for the easy access and free flow of energy, while others require more privacy, stillness, and containment of energy.

When it comes to reading the floor plan of a house, the language of feng shui can be very complex. In this book, I will share with you the basics that are relevant to smudging. As a general rule, there are specific areas and rooms in your home that will benefit from regular smudging, and areas that might not need frequent smudging attention.

These rooms tend to accumulate and store negative energy very quickly:

- Bathroom

- Bedroom

- Home office

- Closets

- Basement

- Garage

- Attic

- Laundry room

- Storage spaces

The bathrooms have the worst reputation in feng shui for a good reason—a considerable amount of clearing old energies goes on there daily. Not many of us know how to help the space do its job so that it does not spread potentially negative energy around the home.

Your bedroom, depending on its design and ratio of windows to doors (we are considering all doors in the bedroom, from closet doors to the in-suite bathroom door to the actual bedroom entrance door), might need more frequent smudging. We release a lot of energy at night, and if you are not in the habit of opening the windows and letting natural light in (or you do not have good ventilation or access to natural light), you need to smudge your bedroom regularly.

The basement, even if it is kept tidy, needs to be cleansed well, as it is the energetic foundation of the home. It is also very important to keep your attic, if you have one, clean and airy, as any heaviness and stagnation in the attic can express itself in an energetic heaviness throughout your home, as well as create blockages in the ability to move forward with clarity.

Generally, the further the room is from the front door, the more it can benefit from smudging. Unless this room or space has a big window or is close to an active back door, it will tend to accumulate stagnant energy. Closets, laundry rooms, home offices, and garages tend to accumulate a lot of heavy or static energy very quickly, so they should be high on your list of priorities in terms of smudging.

The only space in the house that really does not need any smudging is your kitchen—that is, if you have an active kitchen and you cook often! Energetically, the process of cooking is alchemical and purifying for the space, because it uses the powerful natural elements, such as fire, water, wood, and earth. So, if you have a clean and active kitchen, you might not need to smudge it.

Once you fall in love with the process of smudging, you can do what many avid smudging lovers do—have a smudge stick in each of the spaces that needs most clearing. For example, you can have a small smudge stick in a beautiful bowl in your bathroom, a small smudge stick in your bedroom, and then a bigger smudge stick on your living room altar. Or, you can just have one "smudging central," where the smudge stick, the candle, and the bowls are placed. It is definitely best to start with one main altar that displays your smudge stick. Even though I have been smudging for years, I have one altar for all my smudging tools. However, I have also seen homes that have several mini-altars for smudging, and they all look very special and beautiful. This choice will mainly depend on the size of the space, as well as your enthusiasm for smudging. Experiment and see what feels right to you and your home.

The memories stored in your home's corners and walls

We have all heard the expression that walls store memories. These memories are not stored as stories, of course! They are imprinted as specific vibrations, or qualities of energy that our body resonates with and can read at any given time. This is not a conscious process, but rather a subconscious one, similar to your body registering danger as a specific feeling even before you can see the danger.

The main source of energetic pollution in any space—be it a busy street or an office—is the variety of negative human emotions. Ranging from fear and anxiety to boredom, anger, or sadness, we leave imprints wherever we go and rarely even think of being responsible for throwing our negative emotions everywhere.

If you think of your home and recall any unpleasant events that might have happened there, where do you think this energy went? Unless it was consciously transformed into a different quality of energy, it all became trapped in your house's walls, especially in room corners, where energy tends to accumulate and stagnate the most.

Now, go further and think of people who lived there before you—how much have they experienced in this space, and where did all the energy of negative emotions go? It all went to the same place: It got trapped in the walls and any energetically absorbent objects around the home, be it the furniture, the mirrors, or any other decor items. So yes, the previous owners have taken their furniture and their energy with them, but have left you with the burden of all negative emotions still trapped in the walls of the home.

The same applies to happy energy stored in happy homes! This can work for your benefit if you happen to move into a home where people experienced many happy events, and you get a boost of good karma. But surprisingly enough, having too many happy emotions and energies stored in a place can also work against you. Here's a clear example from my feng shui consulting practice. I was consulting for a client that could not sell her house in an area where house sales were booming. The family had already moved to a new house and the old house was listed for sale, cleaned up, and prepped immaculately. Yet, it was still not attracting any serious buyers. Everything was right about the house and the market was good, so it was hard for her to understand why the sale was not happening.

As we were touring the house, I was drawn to a particular area of the living room that was energetically filled with a strange medley of warmth and happiness mixed with deep sadness. I needed to understand more, so I asked for us to spend some time there. After just a little while, my client started opening up and crying, sharing how happy she was in this house, how she did not want to move, and how heavy her heart felt in the new house.

I saw the strong cords that tied the house to her heart and solar plexus areas, and it was clear why the house could not be sold—she would not let it go. Imagine strongly holding onto something while trying to let it go. It just cannot happen. We had to do some energetic clearing, mostly allowing her to experience her emotions and consciously, with full awareness, take them with her to her new home. Clearing the space of those deeply stored imprints allowed the house to be sold very quickly.

This was all easily revealed because we had found the area of the house that was carrying the most intense memories. Every house has energetic spots like the one from the story above that hold heavy amounts of various energetic imprints. You do not need to be a master at feng shui or energetic reading to find them. You just need to have an open mind and allow yourself to listen to your feelings and know when to trust them. This will help open you up to the world of energy and let the energy speak to you.

Smudging clears negativity and pain

We have all experienced painful emotions in our life. Hopefully, our homes act as safe havens in which we can recover, restore the vitality of our energy, and soothe our hearts. While deeply feeling all our emotions is crucial for our healing, it is important to be aware of one fact we often miss: The energy of these painful feelings had to go somewhere after leaving your body. Unless you were out in nature, at a friend's place, or in a therapist's office when expressing your pain, the energy of all the pain and negative emotions you have experienced will still be lingering inside your home. You are still absorbing it and it still influences you, your health, and the overall quality of your life.

Smudging your home regularly will help clear the many layers of negative energy accumulated in your space. If you have been through an especially trying time, I would encourage you to go for a deeper smudging session combined with some space-clearing techniques I will share with you shortly. I would also encourage you to smudge consistently for at least a few weeks after that first deep smudging session.

Negative energy is created and recreated in many ways. While feng shui–wise, negative energy can be caused by specific configurations both in the interior and the exterior

structures of your house, the most common source of negative energy is human emotions. Keeping this in mind and smudging the energy in your home on a regular basis will help you live a healthier, happier, and much more fulfilling life. This is the main reason why shamans use smudging during healing ceremonies—to release and transform the negative energy trapped both in human bodies and in human dwellings.

Use smudging to empower your life direction

The main beauty of smudging is in creating a direct and tangible link with the world of spirit. Smudging, in any of its forms, whether with herbs, woods, or resins, has always been used to open the doors to the Spirit world and ask for guidance, blessings, and protection.

Traditionally, smudging is accompanied by prayers because it is a sacred time in which your vision will not only become clearer, but can also become empowered with certainty. Smudging is often done before long vision quests and sweat lodge purification ceremonies to gently open the wide doors to the Spirit world, as well as to clear one's personal energy of any residues.

Smudging has long been used as a ritual in which to ask the Great Spirit for vision and clarity. You can create smudging rituals to ask for guidance and to empower your life vision and direction. Living life in alignment with your personal vision and purpose is one of the most fulfilling gifts. Smudging rituals can help you achieve this state.

Smudging in order to empower your life direction is not a process that brings quick results, although this can happen, too! It is more of a process of working with very clear intent, using prayer, and then surrendering to receive guidance. Spirits do not work within our human understanding of time, so the best we can do is keep our personal energy and our intention clear. If your intent is to receive directions for your life, smudging is a ritual that will help you get to a place of clarity and gratitude.

Smudging creates peaceful energy

The calm, peaceful energy created by smudging is very nourishing to both you and your home. No matter which tools I use when I smudge my home, whether herbs such as sage,

lavender, or cedar, the sacred palo santo wood, or the frankincense and myrrh resins, the immediate feeling upon inhaling is that of peace. The sense of grounding and relaxation that follows after a deep inhalation has always been there for me from the very beginning.

If you have never smudged your home before, and especially if the process makes you a bit nervous, it might take you a while to feel the sense of peace that smudging brings into your body. Depending on how your space feels, it might require longer smudging sessions, or a combination of smudging and space clearing, but the sense of peace will inevitably follow.

For me, the sense of peace and the delicious slowness of a calmer rhythm comes right at the start of my smudging rituals. No matter what is happening in my life or in my home, once I light the sacred sage, my energy enters a calm and grateful state. Sometimes this calmness can bring emotions to the surface that I was not aware of before. No matter what stages my energy might go through during smudging, the outcome of a smudging ritual is always a more peaceful and gracefully settled energy in both me and my home.

Yes, you can smudge

It is absolutely understandable for you to have doubts about whether you can smudge or if smudging will work for you. Having some resistance to this process can be natural, too. Not many of us grew up in households where ancient traditions were understood, respected, and applied daily. Most of us were never taught about living in tune with nature, let alone about understanding the intricacies of the energy world.

If you have never smudged before, let me suggest the best, most gentle way to enter the world of smudging. Buy a white sage (also called ceremonial sage) smudge stick. These smudge sticks range anywhere from 7 to 12 dollars, depending on their size. If you cannot find a stick locally, you can always buy one online, where there is a wide selection of smudge sticks to choose from. Later, I will teach you how to make your own smudge stick.

Instead of using the whole smudge stick, unwrap your smudge stick (meaning take off all the strings that hold it together), and take just a few branches. Do this carefully, because the branches can break—they are bundled together quite tightly while still fresh so as to take the shape of one big stick when they are dry. If one or two branches break, no worries; you can use them later in your smudging blend. If unwrapping the smudge stick seems daunting, you might even want to start with just one leaf. Or, you can buy a bag of loose leaf white sage. In this case, many small sage branches and leaves are dried individually.

When you have a small branch or a leaf of sage, light a candle, and hold the sage leaf over the flame. Once it takes fire, blow the flame so that it extinguishes and the leaf continues to smolder. Inhale and feel the energy of the plant and fire—how does it feel to you? Do you like the scent? If you burn just one leaf, this smudging will last only a few seconds. You can also try using a full branch, or several leaves, one at a time.

I like this gentle entry into the world of smudging because it gives you time to feel the process, to see how your body responds to it, and to get more acquainted with the actual energy of smudging. Once you have tried this process with the white sage leaves, consider picking up some cedar or juniper branches on your forest walk and bringing them home. Let them dry for a few days and then try the same process with a small leaf from one of those branches. How does that feel to you? Do you like the scent? I also encourage you to try this process with a few branches of dried lavender. The scent is different from that of lavender essential oil, and the energy that smudging dried lavender creates is very sweet and nourishing. Even though comparing herbs is not always easy—or fair!—smudging with lavender to me is somehow similar in its energy to smudging with sweetgrass.

If you allow yourself this slow, gentle experimentation and pure curiosity in getting to know the energy of herbs in smudging, you will gain confidence in your smudging skills. You will begin to develop a more personal relationship with the spirit of specific herbs, and potentially reach a place that feels quite magical and healing.

Rushing into smudging your home with one big smudge stick when you feel nervous and unsure of what you are doing is not a wise way to go. Slow and steady is true for smudging, too! Some people go into active smudging right away, and some take time. Respect the wisdom of your own rhythms and the relationship you are establishing with these sacred plants. We tend to forget that when we smudge, the herbs give their lives in a beautiful fire sacrifice to help us step closer to the life we want to live. Traditionally, it was believed that the smoke from smudging carries your prayers to heaven, and no prayer is ever done in a rush. Consistency, listening to your deep needs and desires, honesty, and authenticity will open up the gateway for you to a lifelong relationship with this powerful process.

I cannot remember the first time I used a smudge stick, but it must have felt very natural because I have not stopped smudging since. Somehow, this ritual connects me to my most cherished childhood memories of spending summers with my grandmother in a tiny, beautiful village by the river. It connects me to the memories of sweet and witty old women collecting and drying herbs, chatting about their lives, cooking, arguing, supporting each other, and laughing.

I was allowed to roam free and endlessly explore everything around me, to climb any tree, to play as long as I wanted, even if late into the night, and to simply follow my own rhythms. Then everything changed, and for quite a few years, the free-spirited child in me had to experience a lot of pain and sadness. When I discovered smudging and started doing it regularly, I felt the same warmth and acceptance as I felt at my grandmother's. I have cried rivers many times after a smudging ritual, but those were tears of relief that came from having touched a place of inner freedom and joy in me.

The more I smudge, the simpler my rituals become. This simplicity does not make smudging less powerful; it makes it more intimate, somehow. It is similar to spending time with a cherished old friend or a wise, loving elder.

I am here to walk with you, step by step, to help you gain confidence in your ability to smudge. Know that the sacred ritual of smudging can help calm a lot of turbulence in your life and soothe your heart. Yes, smudging helps keep the energy in your home clear, but it goes deeper than that. I often smudge to just calm my energy and clear my emotions if there is a lot going on in my life. I also smudge to connect more deeply to myself or to get a sense of the best direction to take in a specific situation. You can definitely learn to smudge, and allow the benefits of smudging to enter your life. After all, this is why you are reading this book. And this is why I am writing it. Yes, you can smudge.

Chapter 3

CHOOSING THE TIME FOR SMUDGING

While there are specific times and cases in which smudging is highly recommended, it is also good to know that there is no wrong or bad time to smudge (though refer to page 12 to see if you should avoid smudging, and page 78 for alternatives). If you have never smudged your house before, now is a good time, because your space has likely accumulated a lot of energetic debris.

Even if you live in the most harmonious home and have the most harmonious energy interactions with your family members, human activity in itself creates quite a bit of energetic residue, similar to the dust accumulated from day-to-day living.

It can be helpful to adopt the practice of smudging your space "for no reason," just as you would drink a cup of tea in the afternoon or in the early morning. Placing beautiful herbs in a teapot, allowing the hot water to steep them, then taking a cherished moment to connect to yourself and your breath is all part of the magic of the tea ritual. Sometimes it is even more about the calming rhythm of the tea ritual than the tea itself!

Once you develop your energetic awareness, you will know exactly when to smudge because there will be a sudden shift in the energy of your space, whether as a result of recent family conflicts, items you brought into your home, or even external activities, like your neighbors going through a big renovation.

As with every ritual and process, there are specific guidelines you can follow in terms of best timing. Specific life situations can call for prompt smudging. In Chapter 8, I will suggest specific smudging rituals you can perform for a variety of life situations, but first, let us explore this idea in detail.

Life situations that smudging can improve

The first step in knowing when to smudge lies in being aware of the negative emotions that have been experienced in your home, either recently or in the past. Smudging can clear the energy after an argument or times of tension between family members. You can smudge your home if you or one of your family members is going through a hard time at work and needs more energetic support. Smudging can also help when you are experiencing heavy emotions such as sadness, anxiety, or fear, or if you just feel depleted and tired most of the time.

Smudging is all about clearing low, stagnant, and plain negative energy. Smudging is also about connecting to the Spirit, the Source, or your own inner guidance, depending on which term you are most comfortable with. This process is important for several reasons. You need to clear the way for a deeper, clearer connection to both your inner guidance and to all people in your life. Living in a house with stagnant and heavy energy is like trying to look into a beautiful garden through dirty windows—you cannot see much! In order to see the full beauty of the garden, you need to clean the dirty windows. Similarly, the negative emotions that tend to accumulate and multiply in your home, if not cleared properly, will distort your view on life.

As we are so deeply connected to the energy of our home—we swim in it and are defined by it—it is important to know that you have the power to shift the energy you are experiencing in yourself and in your surroundings. Clearing the energy around you will facilitate the clearing of the energy inside you. This concept is the whole premise of the ancient art of feng shui and space clearing. A smudging ritual, whether simple or sophisticated, will help you keep the energy in your home and in yourself clear and open.

Challenging situations and events that would require smudging your home include:

- A strong argument
- Abusive behavior
- Having disrespectful guests
- A family member moving out
- Loss of a job
- An accident
- Illness or hospitalization
- Financial loss
- Loss of a relationship

Supporting your energy when you go through challenging stages of life is very important, and having clear energy in your home can provide a huge support. Think of it this way: Never clearing the house energetically is like eating your breakfast from unwashed dinner plates. Would you have your morning breakfast served on a dirty plate? Many homes carry imprints that resemble the grime on these dirty dishes, making it really hard to feel happy

and healthy in these heavily imprinted homes. In terms of feng shui, if the energetic foundation of your home is negative, it is almost impossible to create lasting harmony.

We usually emphasize the importance of clutter clearing for good feng shui. However, clearing the energetic residues and imprints in your home is even more important than clearing the physical clutter. In fact, I often recommend people smudge their home if clutter clearing feels especially challenging. This act can open up the way for trapped energy to flow freely again and make clutter clearing much easier.

Smudging also has applications in happier, brighter instances, such as when you embark on a new journey like moving into a new house, entering a new relationship, or starting a new project. In these cases, smudging creates an energetic opening that allows clear and fresh energy for the success and protection of a future endeavor.

Best times of the day to smudge

Energetically, a day goes through the same cycles as a whole year. The day moves from the depth of night (winter) to early morning (spring) to midday (summer), and finally, evening (autumn). This cycle is one of the reasons why so many people, myself included, love early-morning smudging rituals. There is something both soothing and inspiring about an early-morning smudging ritual; it brings peace, soft joy, and confidence into the unfolding of a brand-new day.

An early-morning smudging, between 5 a.m. and 7 a.m., opens up the way for clear and good energy to come into your life. Smudging in the early morning is energetically similar to smudging during the spring equinox. Smudging early in the morning also helps you clear and center your own energy, thus making you more calm and open to the energies of the day.

Another good window for a smudging ritual is between 11 a.m. and 1 p.m. This time period is considered to have the most potent energy of the day, just like the energy of the summer solstice. I do not recommend performing a large smudging ritual to clear the energy of your house after 5 p.m. because the energy goes into a different cycle and slows down considerably. You can, however, always perform smaller smudging rituals to clear your own energy or to calm a specific room at any time that feels right to you.

Seasonal and moon cycles that add power to smudging rituals

You can amplify the power of your smudging rituals by employing the energy of seasonal and moon cycles. Working with the rhythms of nature brings an energetic boost to your smudging rituals. These cycles do not need to be taken into consideration for your daily or weekly smudging sessions.

The application of seasonal and moon cycles is intended for larger smudging rituals in which you have a very specific intent and know exactly what you are asking for. To understand how to choose the best timing for your ritual, start by clearly defining your intent. Do you intend to clear and let go of something painful and negative, or are you asking for an opening for specific beneficial energy to come into your life?

All smudging rituals are a combination of these two parts, as there is always a clearing of the negative that is done during smudging, as well as a welcoming of positive energy afterward. However, a clearly defined intent for a specific smudging ritual will help direct the energy in the most beneficial way. Aligning the timing of your smudging with your intent will give your ritual even more power.

In a nutshell, you want to choose a seasonal turning point or a moon cycle that corresponds, in its energy, to the energy of your intent. For example, to help release negative energy, whether from a recent romantic breakup or from a big family argument, perform your smudging ceremony during a moon cycle that is waning rather than waxing. A waning moon cycle is the time when the moon is decreasing in size, meaning it is the stage in which the full moon transitions to the new moon. Energetically, this is the best time to let go, release, clear, and purify any emotional residues that are not contributing to your well-being. You can find out the current moon cycle from many online resources, and you can also just look up and watch the moon if it is visible in the sky.

If you want to perform a smudging ceremony to ask for something new—to prepare the room for the new baby, to welcome a new job, or to help clarify your visions and dreams—then the

waxing moon cycle is the one to choose. This is the stage when the moon is increasing in its size and presence in the days between the new moon and the full moon.

Aligning your smudging rituals with seasonal cycles follows a similar logic. You want to match the energy of your intent with the most helpful seasonal energy that can give your smudging ritual more power. For example, the winter solstice has the power to aid solid and aligned energy to envisioning your future, while the summer solstice will help you with bright and fiery energy to quickly manifest your dreams. The energy of the fall equinox helps gracefully and with full awareness release all that is no longer needed in your life and bring clarity to an energetic sorting of various events. The spring equinox is a good time for a smudging ritual to help clear all blockages to welcome new, vibrant, potent energy into your life. In this way, your intent will match and draw from the power of the seasonal energy, given that the spring equinox is a time when life comes bursting forth with immense joy and celebration.

However, you do not have to wait until the spring equinox in order to smudge to welcome new potent energy in your life. Everything in nature is cyclical, so you can always find the desired rhythms in many forms, from the best seasonal and moon cycles to the best days of the week and even times of the day.

My daily smudging ritual

My favorite smudging ritual is the early-morning one. There is something soothing and nourishing in the early-morning smudging; I always feel the energy of love and peace during these short rituals.

This time allows me to connect with any feelings, insights, or dream memories that are still active in my psyche, and to breathe deeply and envision the energy for the day. This ritual takes 3 to 5 minutes at most, and I always use white ceremonial sage. For some reason, this is the only herb that feels right for my energy early mornings, even though during the day I actively use all the options and alternatives I share with you later in the book.

Most of the time I use loose leaves and burn them in a small fireproof bowl. At this time, I do not smudge my space, even though sometimes I might do a brief clearing. I simply sit still in front of the bowl with burning sage and watch the glow. I might use my hand or a smudging fan to direct the smoke toward specific areas of my body. Some mornings, there is the need

for quick movements to clear the energy, and some mornings, I move my hand or the fan very slowly. It all depends on where my energy is at that given moment.

I usually do a few quick smudging sessions to cleanse my own energy during the day, be it in between busy meetings and appointments, or even after a couple hours of work in front of the computer. I smudge every time after I physically clear my home or organize specific areas such as the closets or the office. This process helps the energy get settled after a lot of possibly chaotic movement. Of course, I also do bigger smudging rituals, but not every day!

Another part of my daily ritual is in the evening, especially if I had a busy day and the space feels somehow unsettled. This ritual takes about 5 minutes. I start with sage, then use either lavender or palo santo. I might also use my singing bowl to calm the space before I go to sleep.

Make a ritual into a daily or weekly routine

Rituals bring beauty and meaning to various moments in our lives. They hold special meaning and are mindfully prepared for and celebrated. Rituals often bring the presence of magic into our lives by allowing us to weave in a special relationship, be it with people, nature, or the universe at large. Every culture and every family has its own rituals. They help us strengthen our connections and bring more meaning to our lives.

While I love the energy and the magic of rituals, I have always had a bit of a rebellious streak against making one moment more special than the other. To me, magic is everywhere, constantly unfolding and ever available to those who are ready to see it. In my ideal world, we make every moment a ritual in itself, unique and precious. It is our way to connect and express admiration for the higher power, as well as open our hearts to experience its love and benevolence toward us.

However, I also understand that we live in times that are often devoid of magic, rituals, and deep spiritual connections, so we have to learn our way back to what was once natural to us. Learning the steps for specific rituals, as well as the best times to perform them, is very important for both individual and collective healing. This will allow us to once again infuse life with the power and magic of rituals, so we can accept them as daily routine. And a ritual becoming part of daily routine is a sure sign of a life filled with magic!

I highly encourage you to allow yourself the joy of a regular smudging ritual. As you will see in the following chapter, a basic smudging ritual is easy, takes little time and effort, and brings considerable shifts in both your energy and the energy of your home. You can start by having weekly smudging rituals and then feel when the time is right to do them daily. More often than not, this will happen effortlessly, as your energy will be asking for more of this magical time. The transformative spirit of fire combined with the power of herbs, woods, and resins is waiting for you to use them for better energy and more love, joy, and happiness in your life.

Chapter 4

HOW TO SMUDGE

Learning how to smudge is not difficult. While a smudging ritual for your home can be an elaborate ceremony that takes time and preparation, it is best to start with the basics and build from there. Your house will still benefit from the power of smudging, even if you are following the very basic steps. Smudging is often done before other ceremonies, such as sweats, pow wows, and vision quests, so it is very simple in its nature.

In addition to smudging your house, you can also smudge your personal energy, other people (with their permission, of course!), your land, and specific objects, ranging from personal jewelry to items bought from antique stores and garage sales. Along with white sage, you can also use cedar, juniper, lavender, and other herbs, which are discussed more on page 70.

A simple start is a good start

A simple smudging ritual is a process that consists of several easy steps and very few supplies. Depending on the size of your space, it should not take more than 10 to 25 minutes. Once you understand the basic structure of a smudging ritual and become more comfortable and confident with performing it often, you can start adding various elements of space clearing to your smudging rituals, as well as specific prayers and blessings. I firmly believe that the best start is a simple start. It will help you grow in confidence, as well as increase the enjoyment of this sacred process, thus allowing you to use it regularly with much benefit.

Smudging rituals always have the same structure

All smudging rituals for space clearing have the same structure, which consists of four basic steps. You start with your clear intent for the ceremony, then invoke the help of any energies that can benefit you. The actual smudging of the space comes next, followed by a time of gratitude in completing the ritual with the energy of blessings and protection.

Lighting the candle and **declaring the Intent** for the smudging ceremony always come first. This can be expressed as a simple sentence, such as "My intent for this ritual is to clear the negative energy in my bedroom." You can also express you intent in a prayer; choose what feels most authentic to you. Having a clear intent and declaring it at the beginning of the ceremony, be it silently or out loud, is the first step, and a very important one.

The next step is the act of **Invocation**, in which you invoke the help of spirit in any form that is closest to you (angels, spirit guides, nature beings). This is the step that includes honoring the sacred directions. You light the smudge stick and face each direction, starting by facing the East direction, invoking its power by gently waving the smoke toward it. This act serves as an offering and an energetic opening of the gates for the beneficial energy of each direction to flow toward you and lend its power to your smudging ritual.

Once you have declared your intent and invoked the power of your sacred helpers and all directions, you are ready to go to the next step, which is the process of **Clearing**. This step consists of the actual process of smudging your home. You do this by following a specific pathway chosen beforehand. There are several pathways to choose from when smudging your home; I will share them with you on page 44.

After you have cleared your space, you come back to where you started in order to complete the energetic circle and express your gratitude. This is the last step of asking for **Blessings and Protection** for yourself, your home, or any specific items or situations that were declared in your intent for the ritual. You then extinguish the smudge stick and spend a bit of time by yourself to allow both your energy and the energy in your home to settle.

This is the basic structure of any smudging ritual in four simple steps. There are some details you need to remember, such as having a window open to allow the negative energy to leave and using a bowl under your smudge stick while smudging in order to avoid any ashes falling on your floor. I highly recommend that you smudge yourself before the ceremony in order to clear and center your energy (see page 46 for more on smudging yourself).

My intent here is to clearly emphasize the simplicity of the process itself in order to give you more confidence if smudging is new to you. As you become accustomed to smudging, you can make your rituals more complex by adding some of the additional elements discussed in this book. To start smudging for the first time, though, all you need is to follow this very simple structure and have few supplies at hand.

What you need in order to smudge your home

At the very minimum, all you need is a smudge stick and a candle or charcoal tablet. Yes, it is that simple. In addition, it is helpful to have two bowls, one to catch potential flaming ashes from your smudge stick, and another bowl, a fireproof one, to hold the smudge stick after your smudging ritual. It is best to fill the second bowl with sand, as this is the bowl in which you will place your smudge stick in order to extinguish it. If you do not have sand at hand you can use salt, but please do not use water to extinguish your smudge stick; this can create a harsh quality of energy.

You will also need a designated place to store you smudging supplies. I highly recommend you create a small altar for them. For a more sophisticated smudging ritual, you can add special items connected to your intent, such as crystals or specific visuals, and you can also use sound as an additional space-clearing option. I will guide you through all these steps, from how to create an altar in your home (page 61) to how to use the power of sound in your smudging ritual (page 54).

Having a smudging fan, or a smudging feather, can greatly add to the ritual of smudging, especially if you are using sage bunches instead of the whole smudge stick. You can buy beautiful smudging fans from new age bookstores or native artists' galleries, you can buy them online, and, even better, you can make them yourself. I will show you later in the book how to make a smudging fan.

The main choice to make— bundles or loose herbs

There are two basic ways to smudge your home with sage or other herbs: You can use a smudge stick, or you can burn loose leaves in a fireproof bowl. If you have never done smudging before, the easiest way to go is using a smudge stick or a sage branch rather than burning the leaves in a bowl. Smudging this way is a pretty straightforward process. The only aspect that can make a beginner nervous is the possibility of ashes falling on the floor or any other surfaces during smudging. This is easily taken care of by holding a bowl underneath your smudge stick while smudging.

If you have never done smudging before, I highly recommend using just a few branches— instead of the whole stick—for your smudging ritual. This will help you grow in confidence while progressing to smudging with a full smudge stick. As mentioned before, you can just unbundle a smudge stick and use a few of its branches to smudge your space. Even though I have been smudging for years, I still love to use just a few white sage branches for daily smudging.

The second option, smudging with loose herbs, takes a bit more work, so it is best to work up to it. Just like with burning sacred resins—a process I discuss in depth in the chapter on smudging alternatives—a fireproof container is needed in order to smudge with loose herbs. Even though abalone shells can be used as a container for burning sage leaves, as I am sure you have seen in many beautiful photos, I prefer not to do that, but rather use the abalone shell for displaying sage on my altar. You can also use this iridescent, mystical shell as the bowl to catch falling ashes during smudging.

There is something fragile and pure about the interior of the abalone shell that to me is not a good match for burning objects. It is true that the abalone shell is considered to bring a strong presence of the water element, but I like its presence to remain powerful and pristine, rather than burned. However, many people do use the shell to burn sage leaves, so it is all a matter of personal choice and preference. Traditionally, there are Native tribes that use the abalone shell in which to burn leaves, and there are also tribes that prefer not to use the shell for smudging. So, the choice is truly up to you. The abalone shell is not fireproof, so while you can use it to burn a couple of loose leaves, it is best not to use it for longer ceremonies, when you have to smudge a whole house.

If you decide to smudge your house with loose sage or other herbs, first you need to have a bit of sand at the base of your fireproof container before adding a lit charcoal tablet. Next, place the sage leaves on top of the charcoal tablet and wave the smoke produced by them into the air. You can use your hand to distribute the smoke, or you can use a feather or a smudging fan.

Just like with smudge sticks, the loose leaf white sage or smudging blends that include a variety of herbs such as lavender, cedar, juniper, rose petals, and other herbs, can be purchased online. You can also easily make them yourself. Depending on what is available in your area, you can pick cedar, juniper, lavender, sage, rosemary, and rose petals, dry them, and make a smudging blend. It is best to start with a simple blend with just one or two herbs, then add more options in time.

Three smudging factors to know

There are three important factors you need to consider when planning a smudging ritual to cleanse the energy in your house.

1. Intent and timing. Decide on the best timing—day of the week, time of the day, season, or moon cycle—for your ritual, as well as the length of time you can devote to the session. You might have only 5 minutes, or you might have an hour to cleanse your whole house. Find a time when you will not be disturbed. Be sure to have complete clarity on both your intent and your timing before you start the session. Also, ensure that you will have uninterrupted time for smudging, as well as easy access to all spaces you want to smudge.

2. Your energy. It is important to spend a bit of time focusing on your energy before you proceed with the smudging of your space. Take a few deep breaths and stretch or move your body to feel more connected to yourself. Maybe you need to take a shower or have a cup of tea before you start. Decide what your energy needs in order to feel more present. The more centered and present you are, the more powerful your smudging session will be. Of course, once you start smudging, you will start feeling noticeable shifts in your energy.

3. The process and supplies. The basic process of smudging is very easy to master, and the powerful effects stack up in time with regular use. You can also choose to amplify the energy of your smudging ritual by adding various tools such as the power of sound, water, or various

visuals, as discussed in detail in the next chapters. Be sure to play and add more elements in time as you get more comfortable with your specific ritual. As mentioned before, the process of smudging can be done with a whole smudge stick or with just a couple of sage branches, which is my preferred choice for daily smudging rituals. It is always recommended to have at least one window open so that the lower energy can be released.

Your step-by-step smudging ritual

The main goal of smudging is to purify and uplift the energy of any given space, be it an indoor or an outdoor space. You can also smudge to purify your personal energy, as well as a variety of objects around you.

Let us start with the smudging ritual for your home first. You will need:

- ❁ A smudge stick
- ❁ A bowl filled with sand

- Another bowl to catch ashes

- A candle

1. Place all items on a flat surface away from drafts and farther away from both the front and the back door. Ideally, you will find a place close to the center of your home, called the heart of your home in feng shui. If you already have a home altar, it can be beautiful to start your home smudging ritual there. If you do not have an altar but would like to create one, the instructions are there for you on page 61.

2. Remove all rings and bracelets, unless they have specific meaning for you in this ritual and you know how to cleanse their energy regularly.

3. Be sure you have a window open to allow negative energy to be released.

4. Light a candle and spend a minute or two focusing on your intent for the smudging ritual. It can be as general as "I want to clear the negative energy in my home" or as specific as "I want to release all the sadness lingering in my bedroom after my recent breakup." Do not forget to take deep belly breaths while focusing on your intent. This first step is necessary in declaring your intent.

5. Next comes the step of invocation, when you invoke the presence of energies that can help with your request. Hold the smudge stick in your dominant hand and place its tip over the candle in order to light it. Once ignited, let it burn for a few seconds, and then extinguish the flame at the tip of your smudge stick so that it smolders. Gently move the smudge stick to wave the smoke into the air around you. Invoke the help of your spirit guides, angels, and helpers, or say a prayer to the Great Spirit. Honor each of the sacred directions by facing them one by one and gently waving the smoke toward each one as you face it.

6. To proceed to the actual step of clearing, keep holding the smudge stick in your dominant hand and hold a bowl underneath the smudge stick into order to catch occasional ashes. Start smudging your home by following a specific route that feels best for you and your home. No matter which way you choose, the smudging is always done in a clockwise manner. Here are the options to choose from:

- You can start by facing the direction of East right where you are, at your altar or at the center of your home, and start smudging your home, room by room, from there.

- You can go to your front door and start smudging your house from there.

🔥 You can start by smudging the room farthest from the front door, which tends to accumulate many layers of stagnant energy.

To purify the energy in your home by smudging, slowly wave the smudge stick into the air, starting at the higher levels closer to the ceiling and then smudging the levels closer to the floor. You might notice that you spend more time in some spaces than in others; just trust the energy and follow it. Be sure to spend more time on all room corners, as this is where the negative energy accumulates and stagnates.

Do not forget to use your bowl to catch ashes. Be especially mindful around upholstered furniture, various fabrics such as curtains and bedding, and rugs. You might need to relight your smudge stick a couple times during smudging; this is absolutely normal, especially with thinner smudge sticks. To relight your smudge stick, just return to your altar and hold the tip of your smudge stick over the candle flame till it ignites. Once ignited, let it burn for a few seconds, and then extinguish the flame so that it smolders again.

Proceed from room to room, wall to wall, corner to corner, until you come back to the place where you started. If you feel you need to go another round, trust your feelings and smudge one more time. If you feel a specific corner is asking for your attention, go there.

7. Once you are finished, gently extinguish your smudge stick in the sand-filled fireproof bowl and leave it there for a while. This is the stage of blessings and protection, when you consecrate the energy in your space with specific qualities, such as peace, joy, or just vibrant fresh energy to welcome new beginnings. You can also ask for protection for your home.

The last step is important because it gives a specific direction to the newly cleared energy. It is like imprinting it with your intent and defining a clear container for it. No ceremony is ever complete without expressing gratitude, so be sure you find your own way to express this energy. It can be a silent prayer, words spoken out loud, or even a dance or a song! Find the most authentic way for you to express heartfelt gratitude for the ability to use these sacred herbs to purify your home.

This act concludes the actual smudging ritual. However, for best results, please do not rush into other activities. I highly suggest you spend some time in silence recalling your experience with smudging and writing down any specific memories, impressions, or feelings you had during this process. You do not have to write things down every time you smudge your home, only when you have helpful insights that are worth exploring further. Be sure you have this uninterrupted time after the smudging, even if it's just 5 minutes. The energy right after smudging feels really special and powerful; it has a quality of calm strength to it. You will want to benefit from it and fully soak in this energy.

If you have done a bigger smudging ritual, you might have a lot of sage ashes left that you can use in your garden as nutrients for plants.

As you can see, the process of smudging is easy. It only takes some simple precautions to avoid dropping any flaming ashes on your floor, furniture, or bedding. I have been smudging for years without burning any furniture (and I rarely use the bowl), so I know you can get really skilled at it. To begin with, though, use the bowl to be safe, and do your smudging at a slow pace.

Now, let us also be sure you know how to smudge your own energy. This is a joyful, helpful, and quick process to do for yourself daily, either in the morning or in the evening. Smudging your energy is also recommended before you start a house-smudging ritual.

You can (and should!) smudge yourself

Smudging yourself before smudging your home—or smudging others—is always a good idea. You can also smudge yourself when you do not plan to smudge your home. For example, I might do a quick smudge before going out if I feel a bit nervous about the trip; the act of smudging always grounds and centers me. I also smudge if I have visited many places during the day and feel I have picked up a lot of energies that are not beneficial to me. Smudging has become a daily ritual for me that I love and even look forward to. No matter what works best for you—regular daily smudging or quick smudging when most needed—this is really easy and hardly takes a minute or two.

If you do not have your altar set up, reach for your candle and the bowl with smudge stick, and place them somewhere convenient for the next couple minutes. Be sure to find a place away from upholstered furniture or fabrics so that your mind can be at ease if a couple ashes fall on the floor.

Here is how to smudge yourself.

1. Light the candle.

2. Take a couple of deep breaths. Move your body if needed. You might stretch your arms or your neck, or maybe do a bit of Qigong type shaking (freely shaking your body to release stagnant energy).

3. Ground and center yourself. To ground your energy, imagine a cord extending from your navel to a point about 12 inches beneath your feet. You can also extend it all the way to the earth's core.

4. Express your intent or say a simple prayer. A short sentence such as "I intend my energy to be cleansed" can work as a declaration of intent. A longer prayer can be beautiful, too. It is the clarity of your intent that matters the most, so choose what feels most authentic to you in the moment.

5. Light the smudge stick or the sage branch from the candle, then extinguish the flame.

6. Start by waving the smoke around your heart area, then move to your third eye (the spot between your eyebrows), then to the space above your head. Wave the smoke in gentle movements with your hand or your smudging fan or feather. Spread the smoke above you, then in front and behind you, and finally around the sides of your head and shoulders. You are clearing your aura, your personal energy field, so you do not need to be too close to your body. Holding the smudge stick 10 to 15 inches away is just fine.

7. Proceed to smudge the front of your body, waving the smoke around your throat, heart, shoulders, arms, hands and palms, abdomen, sacral area, legs, and feet.

8. Spend some time waving smoke around the soles of your feet.

9. Repeat the same process at your back as best you can, then your sides.

Do not worry that you can't fully reach your back; nobody can. The smoke knows where to go if your intention is clear.

10. You can repeat the same actions by continuing to smudge one more round—from your feet back to the area above your head.

Done! Just as with smudging your home, there are several ways to smudge your personal energy. Sometimes I see recommendations to smudge your energy by starting at your feet. This approach does not resonate with me, so I smudge my energy in the way that feels the most harmonious to me. I start by clearing the energy around my heart, then go higher up to clear the energy around my head to open my connection to higher wisdom. Then when these two centers—the heart and the head—are more clear, I take this energy to help clear the rest of my body and ground it in my feet.

Depending on where your energy is at any given moment, you might find yourself smudging differently on different days or at different times of the day. For example, you might feel like you need to start by smudging your feet if you need more grounding, or start at the head if you had a busy day at work and are smudging yourself in the evening.

My favorite time to smudge is early morning, before the dawn of a new day. Smudging in the evening has power, too, as it helps release energies you might have picked up during the day that are not beneficial for your well-being. Some energies are not necessarily negative, just not needed. A good exercise to do when smudging in the evening is to close your eyes and visualize all connections you had with people during the day as specific cords attached to you. If you are particularly sensitive to energy, this will not be just a visualization exercise for you, as you might actually see the energetic threads still connecting you to those you interacted with during the day.

Some of these cords will be much thicker and have more charged energy than others. It all depends on the nature of that specific interaction. Focus on seeing a bright light in your heart and expand this energy to as big a size as needed in order to dissolve all the cords. This will keep your energy clear and avoid the unnecessary agitation or confusion we often feel while interacting with too many people during the day. Taking time to regularly clear our energy contributes to better quality relationships, so do not be concerned about dissolving cords even with those you love—you are not cutting them, you are just allowing fresh, new energy to enter all your interactions.

As always, I offer you what I know, and I encourage you to experiment and find what feels best for you. There is no wrong sequence in smudging yourself, just as there is no wrong sequence in smudging your house. As long as a specific sequence feels calm and peaceful to you, it is right for you.

How to smudge others

There are many circumstances when smudging others is helpful. If your partner or your child came home from visiting a busy place and you feel them carrying a lot of heavy energy (this can be expressed in their emotions or behavior), it is always a good idea to smudge. Of course, this implies having clear agreements beforehand about the acceptability of smudging. This also means asking for permission to smudge, every single time! You cannot just light the smudge stick and proceed to smudge your partner because they clearly liked it the last time, or because you see they are carrying negative attachments. Please wait for a clear "Yes" before smudging anyone.

My daughter makes the happiest face ever when I smudge. She has been in love with the scent of sage since she was a tiny little muffin! Still, even though she adores it when I do even a quick smudge around the house, I always ask for permission if I can smudge her own energy.

If you teach a spiritual workshop, lead a circle, or provide a more alternative healing service, you might also offer your students or clients to be smudged before the class or the session.

The best way to smudge someone is for them to be standing, not seated, if possible. Ask them to stand with their feet slightly apart and arms slightly spread. Of course, if the person is way taller than you, or they feel sick and need to sit down, smudging will still work. Ask for their permission and hold a clear intent or a prayer in your heart. You can also say it out loud, depending on what feels best to you. Tell them to take a couple of deep breaths and relax. Breathe with them and focus on visualizing a source of bright light in your heart. You can see it as a golden shining sun. Take a couple of seconds to see this light growing and enveloping you completely in its orbit.

Start at the head of the person—12 to 15 inches away from their head—and continue by following the same steps as you do when you smudge yourself. I find it feels more natural when smudging others to start by smudging their side rather than their front. If you start by facing the person directly, this position might feel intimidating or even confrontational on a subconscious level. If you do feel like starting to smudge by first smudging the front of their body, please be sure your body is on a bit of an angle from them rather than directly facing them, as this psychologically feels less threatening.

Always keep the smudge stick at least 12 to 15 inches away from their body. Do not forget to smudge the soles of their feet and the palms of their hands. After you have finished, take a moment to sense if you need to repeat the smudging steps, or if the energy feels better already. If the person is breathing deeper, this is a sign that their energy is being cleared. Connect and smile, then spend a short moment in silence. Extinguish the smudge stick and place it in your bowl with sand.

Smudging specific objects

The cleansing and purifying power of fire and herbs can also be used to smudge and conse-crate specific objects. The items I smudge the most are my jewelry pieces, my crystals, and sometimes, specific pieces of clothing. I might also smudge my Buddha statues or any decor items that I feel can benefit from some cleansing, such as my wall tapestry or beautiful dream catchers.

With both jewelry pieces and clothing items, I prefer to keep them in one hand so that the smoke can freely flow through them in all directions, and smudge them with the smudge stick held in my other hand. You can also just lay them on a surface to smudge them, but it is way more powerful to have them freely floating in the waves of smoke. For example, I might hold my turquoise necklace in my left hand and gently wave the smudge stick around it with my right hand. I might move the hand holding the jewelry, too, or I might hold it still and just move the hand with the smudge stick. It is good to do this for about a minute or so, and not rush. Afterward, you can gently place your jewelry piece on your altar or on any other special surface before putting it back to its usual place.

The same steps work with clothing. You just have to be way more careful and hold the hand with the smudge stick much farther from your clothing, unlike when smudging jewelry, when you can keep it very close. Why would you choose to smudge your clothing? Good question! I have girlfriends who will smudge their clothing before going on a date (lightly, of course!). I also know people who will smudge specific items they would take to a job interview. I smudge some pieces of clothing that are not washable, or that are washed very rarely, just so as to give them a boost of fresh energy.

It is also very important to smudge items you brought home from antique sales, secondhand stores, or garage sales. You never know what energies they hold, so it is always best to be on the safe side and clear the energy of any given object that has been owned by someone else for a long time. Items hold energies and memories, and many of them can be quite painful, so it is best to clear it all out before allowing it entry into your home.

Chapter 5

SMUDGING ADD-ONS

There are a wide variety of tools that you can add to your smudging rituals in order to enhance them and to focus their energy in a specific direction.

Use these simple space-clearing tools

Now that you know that most spaces carry heavy energetic residues, you will understand why all ancient cultures had at least one form of space clearing they constantly used. These various forms of space clearing were developed based on many experiences and gifts from the elders. The wisdom of space clearing is simple—each space needs to be cleared, energetically, on a regular basis. The ways of space clearing are many, and you can choose what works best for you.

Essentially, an energetic space-clearing session uses at least one, and more often several, natural elements for their cleansing powers. Smudging, which uses the element of Fire, is one of the most powerful forms of space clearing. There are also various ways of using the Water element, from sprinkling the whole space with blessed water to placing bowls of pure water on your altar and changing them daily. Aerating the space often and placing specific crystals are also among the most common ways to space clear by using the Air and Earth elements.

Using the natural power of salt—the Earth element—to absorb negative energy is common in many cultures. Simple steps such as sprinkling salt in the corners of any room, or even the whole floor of a specific room, and leaving it there overnight, can do a good job of cleansing some negativity. Collect it the next day either with your vacuum or with the broom, and discard it outside the home.

There are many ways to clear energy that you can include in your smudging and space-clearing sessions. Again, approach it with a spirit of curiosity and see how your energy responds to it. The power of visualization is also often used in smudging. You can visualize the specific results you want to achieve, such as having more energy for your projects or more harmony in your romantic relationship. You can also choose general visualizations of a happy home filled with light and laughter.

Add any of these techniques to your own smudging rituals. In time, you can add elements stemming from your own joyful creativity, as well as from the unique voice of your house.

Five forms of sound to clear energy

Using the power of sound is one of the most popular and effective ways of space clearing. Sound is a primordial power, and some believe that the vibration of sound created the whole universe. Sound is able to both create and destroy matter; we know that all matter vibrates with sound.

There are five easy ways to help you add the power of sound to your smudging and space-clearing sessions. You can use:

1. Recorded music

2. Singing bowls

3. Drumming

4. Clapping

5. Your voice

Music: The easiest way to shift the energy in your space while smudging or space clearing is to play powerful music in a clean room with open windows, ideally during the daytime. The music has to be fairly loud, and it has to be powerful and meaningful to you. You can choose music with meaningful words (be it Enya, Loreena McKennitt, or Gregorian chants), or music with no words that evokes specific emotions and feelings in you. You can calm the space, energize, or balance it with specific tones. Words are very powerful, so be mindful of not playing music that you might like but has sad words (as is the case with most popular songs). Be mindful of the energy you are allowing into your space.

Another option is to use **singing bowls**. This might sound intimidating if you have never used them, but I assure you it is not only easy, but fascinating, too. Basically, the singing bowls, also called Tibetan bowls, are a type of bell in the form of a bowl. Traditionally, they are made from seven metals and often have various designs inscribed on them. I love using my big singing bowl and I do it intuitively; nobody taught me how to use it. Choosing a singing bowl is a very personal process. You can find a variety of different sizes of singing bowls to suit different budgets. You can explore many singing bowls for sale online, mainly to get an idea of their prices, variety, and value. I highly suggest, though, that you choose your singing bowl in person by trying several different ones and spending some time with their sound, look, and feel.

What makes a singing bowl so powerful? The sound of the singing bowl has the ability to bring everything back to a center of stillness. Try one at a store or listen to a recording of a singing bowl and feel how your energy responds to its specific sound. Another great benefit of using singing bowls for energetic clearing comes from one of my crystal healing teachers—she uses the sound of crystal bowls to cleanse all her crystals at once. She places the crystals on a large table and then uses the sound of singing bowls to clear the energy of all crystals without spending time on clearing each crystal separately.

Drumming has become very popular and many of us have beautiful drums at home, or even make our own drums. I have a beautifully designed drum with a flower-like mandala of women's bodies, and I love playing it when the energy feels right. Drumming evokes a primordial, natural, and powerfully healing rhythm that we are all aligned with. It has the ability to clear and balance all the misaligned energies in any given space. It is the sound of a universal heartbeat that aligns everything with its heart power. If you have never done any drumming, allow yourself to explore it with a beginner's mind, or childlike curiosity. The needed rhythm will make itself known to you once you start drumming; all you need is the willingness to listen. Nobody taught me how to drum, either. I just fell in love with it and use it when my space or my energy need it.

Clapping to clear your space might sound silly to some. However, the simple action of clapping your hands is very complex in energetic terms. Think of the namaskar mudra, the most popular mudra in all yoga classes, when we connect the palms of our hands at the heart level. This gesture helps enhance the alignment of all energetic currents in your body, from aligning your chakras to balancing the right and left sides of your body. When we clap for the purpose of space clearing, we facilitate a clearing and an alignment of energies. We use the energy activated in our hands to clear any out-of-alignment, stuck, and stagnant energies in our environment.

As stagnant energy tends to accumulate in corners, clapping the corners of any room is a well-accepted form of space clearing. You start at the lower level, closer to the floor, and proceed to the higher level, closer to the ceiling. Try it for yourself and notice the clearly distinct difference in sound while you clap from lower to higher levels. While you are clapping closer to the floor level, where the lower energy accumulates, the sound will be muddy and dull, and you will fell a sticky quality of energy in your hands. Keep clapping as you slightly move up your hands and the energy will change, then continue going higher up. To clearly feel the difference, clap once or twice at the lower level of any corner and then move your hands higher up and clap again. You will be surprised at the difference in the sound and the feel of the energy there.

Human voice is very powerful, and we have yet to learn about all the secrets of manifestation by using our voice. The topic of the sound of your voice is a very complex one. It starts with the resonance of your voice and goes all the way to your ability to speak your truth and stand up for what you believe in. While this is fascinating for me to explore, here I have to stay with the topic of using your own voice in order to shift and clear the energy in your space. This might sound strange, but I assure you that it works. I understand that this specific way can be intimidating and most probably not something you will be willing to try first. Unless you are brave enough to sound like a shaman right away!

To experiment with using the sound of your voice for space clearing, I suggest choosing a time when you can be alone and not concerned with being interrupted. Try any sounds that come to you when you intend to clear the energy in a specific room. You will be fascinated by what comes up. It is a dialogue and a revealing of layers and layers of energy. You might have watched this process during shamanic healing or clearing ceremonies; it is very powerful. I acknowledge that the act of using your voice for space clearing can feel uncomfortable to you. There is no rush. Start by giving the idea some time. I have yet to master the courage to regularly use my voice for the purpose of clearing energy, but I do it sometimes. The benefits of this form of space clearing goes way beyond just clearing your space—it opens up a strong energetic channel in your own body that brings more empowerment and courage into your daily life.

These five ways of using the power of sound can be used with smudging in any combination you like the most. Maybe it is the combination of drumming, smudging, and then using singing bowls (my favorite); maybe it is your voice, clapping, and then smudging; or maybe it is simply a ritual where you play beautiful music and slowly smudge your space. I am giving you all these powerful options for a pick-and-choose process to add to your own ritual. Allow your own energy to make the choice, a choice that works best for you and your home.

Clarify your intent for a powerful smudging ritual

The power of consciously using the energy of intent is a fascinating topic and a process that not many of us are yet good at. It certainly did not come easy to me. I have to admit I rarely

used to clarify my intent or consciously work with it, unless it was plainly obvious and energetically strong.

My lessons on the necessity of using my intent came all stacked in a row, quickly, unexpectedly, and some with a power that frightened me. It all started when a healer I was working with hung up the phone on me saying that she could not work with someone who had no clear intent for the desired outcome. I was quite shocked. Little did I know that this experience was nothing compared to a way bigger lesson that life had in store for me. After that phone call, I started exploring everything about the energy of intent—from what it is and how to use it to how to actually clarify your intent. I read books about it, I discussed it with close friends, I journaled about it, and I was still very confused. My head was brimming with questions. Isn't it egotistic for me to impose my own intent over the Divine will? How do I know the difference between following the Divine will and living with intent? How do I know if they are truly aligned and, if they are aligned, why do I need to have a clear intent when I could just trust and surrender to the Divine will? How do I know the highest intent in any given moment? My mind was going in circles, which is never a good sign, so I decided to just focus my intent on understanding intent. As simple as that.

Shortly afterward, while traveling in Australia with a group of friends, I had an experience that shook me to the core. There were no questions left in my mind about the use of intent; every cell in my body got the lesson.

We were driving in a car in the most beautiful Australian hinterland, and I felt that our driver started behaving strangely. Because I work with energy, I know the difference between shifts in my emotional energy versus a pure energetic reading telling me something is not right. While the driver's behavior was becoming more and more erratic, I saw, in my mind's eye, three very clear possible outcomes for the situation we were in: We could die because the person at the wheel was becoming more and more reckless, we could end up in the hospital, or we could reach our destination alive and well. The last outcome, reaching our destination unharmed, seemed least probable at that moment. I felt a very strong sense of fear, as well as urgency to make a decision. My mind went through a million possible solutions to prevent the worst outcome. Then, a sudden sense of stillness came over me and I saw the face of the shaman I spent time with the day before. He was looking straight into my face, saying, "You have to learn to live with very clear intent. You have to know what you want. What is your intent right here, right now? What do you want the most?"

And I said, "I want to live, of course! I do not want to die here." To which he simply and sternly replied, "Then focus on it with every bit of your energy. Focus your intent fully on

what you want." So I focused all my energy on clearly seeing myself back in my hotel room. I was making it all very real and sensory—opening the room door, feeling my feet touching the carpeted floor, hearing the squeaky sound the balcony door makes when I slowly opened it to let the ocean breeze in. I was visualizing it again and again and again with more and more power, focusing all my personal energy on the desired outcome, while our driver was going in and out of control on the beautifully winding and narrow road.

This experience lasted for almost an hour, and I could literally see how my energy strengthened the desired outcome when I was fully focused on it. I also saw how the possibility of the other two outcomes became stronger when I drifted away into feeling fear instead of focusing on my clear intent.

"Have a very clear intent, know what you want in any given moment, and trust the Mother," the shaman kept repeating to me, as if aware of my long struggle with personal intent versus the Divine will. That was the missing piece for me, the piece that was somehow limiting me from living my life fully. While I was really good at asking for guidance and trusting it most of the time, I was definitely stunted in the expression of my own intent. In that moment, I not only understood; I also clearly felt, in every cell of my being, how a clear intent goes along with trust and surrender, and I knew I could do both.

So, what does the notion of intent—or the lack of it—have to do with smudging? Quite a lot. While the most basic and practical use of smudging is to clear negative energy, as well as to purify the air of harmful bacteria, the deeper and more powerful use of smudging is to establish a direct connection to Spirit and to your own innermost Self.

Declaring your clear intent before each session is an empowering act of focus, trust, and surrender. Your smudging ritual becomes a form of prayer, an expression of love, and a process of co-creation. From this place of devotion and trust in higher guidance, your intent takes on tremendous power. Creating an altar dedicated to a specific intent can be very helpful, and so is choosing a specific smudging ritual for your intent. You will find many examples of using clear intent for various smudging rituals in the last chapter of this book.

I would like to encourage you to take a few moments each morning to clarify your intent. What is your intent for the day? For any specific actions you are taking? What energies do you want to invite into your life? Clarifying our intent and expressing it in meaningful rituals speeds up the manifestation of our dreams. It reminds us that magic is always here, patiently waiting for us to step up as conscious co-creators of life.

Prayers and blessings to use with smudging rituals

Prayers and blessings always add more meaning, beauty, and power to smudging rituals. Some of us pray a lot and are comfortable with saying prayers many times throughout the day, while for some of us, saying prayers and blessings can be new and feel a bit uncomfortable. You do not necessarily have to incorporate them in your smudging if this does not feel natural to you yet. It is important, however, to express the energy of gratitude in any way that feels right to you. It is also very important to be clear with your intent for the smudging ritual and to ask the higher power—in any way that is closest to your heart—for help in clearing any negativity.

I do not use different prayers for different smudging rituals. My prayer of gratitude is always the same. I pray to the Great Spirit and the Divine Mother, and I use the words that feel most genuine to me in that moment. I declare my intent for the specific ceremony, I ask for help, and I always bless my space while smudging. I start any of my smudging rituals by giving thanks for the ability to use the alchemy of fire and all these powerful herbs, woods, and resins, and I finish by giving thanks, too.

Saying blessings while smudging is a beautiful way to infuse your space with very gracious and happy energy. It is especially powerful to say these blessings when you smudge yourself or someone else. You can say them out loud or silently, depending on what the moment calls for. Here is a simple blessing I adapted from various prayers. Feel free to recreate it the way it feels best to you. It is used for smudging personal energy, either yours or someone else's. You bless each part of the body as you smudge it, clearing it, and wishing for its best expression in the world.

May my heart be cleansed
To always be soft and open

May my throat be cleansed
To always speak my truth

May my eyes be cleansed
To always see beauty and kindness

May my mind be cleansed
To always know the truth

May my arms and hands be cleansed
To express the Spirit's work and beauty

May my feet be cleansed
To work my path in strength

May my whole being be cleansed
To feel the love of the Great Spirit

Washed clean, renewed, and rejuvenated,
I give thanks for my life

When smudging someone else and using this blessing, be sure to substitute "my" for "your." Feel free to add any words that come to mind when blessing your energy or someone else's. Trust that the moment is bringing you the expression you need the most.

How to make your own altar

There is so much that can be written about altars. It is a fascinating topic and a process that all cultures around the world use. In its essence, an altar is a special place designated to honoring the Spirit world, God, the Universe, and Life in all its expressions. It is a place for offerings to the Spirit world and expressing gratitude for your life. An altar is also a way of expressing your way, or your presence, in this web of life. It is an energetic declaration that you are participating in the mystery of life; it is a mini-universe of your own making. I have been creating altars in my home for a long time. It is hard for me not to create one if I see a good space for it. At the moment, I have three main altars and then a couple smaller ones. My energy is naturally inclined to create altars everywhere.

While the concept of creating an altar can sound intimidating, I am sure you have done it at least a couple times in your life! On an unconscious level, every time you set beautiful pieces on your fireplace mantel, you are creating an altar. Any time you play with different decor solutions until you find just the right one for your coffee table, you are creating an altar. We are already creating little altars in our home each and every time we group objects we love to create a happier home. Whether on a mantelpiece, a bedside table, or the coffee table, you have worked with this energy before, just perhaps not consciously.

Altars can be generic, as well as specific when dedicated to a clear intent, whether it's the intent for better health or to welcome love. All altars need energetic movement and care, so ideally, you would care for your altar daily. This can be as simple as lighting a candle and slightly rearranging the items on your altar. It is the outpouring of your energy that matters the most in creating and maintaining a powerful home altar.

The key here is conscious creation. When you are creating an altar in your home with a specific intent in mind, and you lovingly care for it, your altar starts accumulating the power needed to support your dreams. If the concept of an altar is very new to you, I will give you the most important guidelines for creating an altar. And if you are a lover of creating altars, you might find some helpful ideas here, too.

The main thing to know about creating an altar in your home is that there are no set, specific, or rigid rules for their creation. Your home altar is an expression of your own intent and unique personal energy, so the more empowered you are in creating your own altar, the more power gets infused in it.

While there are no strict rules for creating an altar, there are guidelines you can follow to make the altar creation easier for you. This can be very helpful if you have never made an altar before. Here are some basic steps to consider when creating an altar for your home.

1. The area. Decide on the area and the surface of the altar. Will the bedside table work best for your altar? How about that open area on your bookshelf? Maybe the windowsill is calling for it? Is your fireplace mantel ready for a makeover?

2. The topic. Feel or decide on the main theme of your altar. Do you need more clarity in your life? More energy and vitality? Do you want to express and attract more love? The love altar, the gratitude altar, and the connection and clarity altar are the most popular altar themes. You can also just let the topic evolve in the process of creating an altar.

3. Energy enhancers and tools. It is always helpful to contain the energy of your altar, as this makes its energy more powerful. In order to do that, you can place a beautiful piece of fabric or a tray as an energetic foundation for your altar. Crystals and stones always bring powerful presence to any altar, and so do candles, flowers, specific images and sculptures, and special items from your nature walks.

I highly encourage you to create an altar to hold all your space-clearing and smudging tools. You want to have a respectful, beautiful place to rest your smudge stick or your sage branches between smudging sessions. You can also display on your altar any items connected to your smudging and space-clearing sessions, such as a smudging fan or feather, an abalone shell or a bowl, and a candle. Having them on an altar will not only infuse them with more power, but will also remind you to smudge more often.

Crystals for a smudging altar and for your home

Crystals and stones are awesome additions to any altar and any home. All crystals bring powerful energy from deep within the earth, a quality of energy we dearly need for our health and well-being. It is important to know, though, that many crystals are colored, treated, and irradiated, especially some of the most popular ones such as smoky crystals and citrine. These treatments can change their properties and diminish the potency, so it is always best to buy raw, natural crystals. There are also many crystals that are synthetically made, and even though this caution applies mostly when buying jewelry, it can also apply to little carvings or amulets you might buy for your home altar.

The best way to benefit from the energy of crystals and stones is to have them in their raw form, meaning a cluster, a point, or a geode. If this is not possible for you, choose crystals in their tumbled form. It is important to cleanse your crystals when you first bring them home. Just like people, crystals can benefit from a good smudging session to clear all vibrations from places they have been before—unless you get one straight from the mine, which would have incredibly powerful energy right from the womb of Mother Earth!

Any natural crystal or stone will be a joy to have on your smudging and space-clearing altar, so in choosing crystals, it is good to be guided by your intuition. Some of the most helpful crystals to place on your smudging and space-clearing altar are amethyst, black tourmaline, malachite, carnelian, and clear quartz. If you are drawn to a specific crystal not listed here, surely bring it home to your altar. The crystals below are just suggestions based on their properties, as well as the ease of obtaining them.

Amethyst. Ranging in color from pale lavender to deep purple, amethyst is one of the most popular stones in crystal healing and feng shui. Amethyst is stunningly powerful and beautiful, easily available, and very affordable. Choose an amethyst cluster or geode if you need more peace and calm in your life, long for a deeper spiritual connection, and desire a slower, clearer, and nurturing energy in your home. Amethyst is also very helpful if you are trying to meditate, learn more about yourself, or connect to a higher source of wisdom.

Black Tourmaline. As a defender and protector of good energy, black tourmaline is an excellent addition to your smudging and cleansing altar. It aids its energy to the already powerful cleansing energy accumulating around your altar. To benefit more from the strong protective and cleansing energies of black tourmaline, you can wear it as jewelry, such as bracelets

or pendants. It is also a good idea, feng shui–wise, to display a couple of black tourmaline stones close to the main entrance of your home.

Carnelian. The inspiring carnelian combines the energies of both fire and earth elements, making it an ideal stone for your altar. Ranging in color from gentle yellow to deep fiery orange, carnelian also promotes grounding, confidence, and a joyful outlook in life. This is a stone that can lend you the energy to break through limiting beliefs and access your power in a calm and wise way.

Clear Quartz. The crystal of all crystals, clear quartz will absorb and carry any energy you program it with, so you can program it to carry specific energy on your altar. You can display a clear quartz crystal cluster, a quartz wand, or just a couple of tumbled stones. The beauty of having several clear quartz tumbled stones on your smudging altar is that you can alternate them and carry one or two with you in your pocket or purse when you need to feel more protected. If you have not used your tumbled clear quartz for a while, do not forget to cleanse it before placing it back on your altar. This should be really easy, as all your smudging supplies will be right there, on the altar!

Malachite. The energy of this greener-than-green stone is cleansing and energizing. It can bring deep transformations, help open your heart, and encourage you to speak up and be more brave. It is wonderful to display a malachite on your smudging and cleansing altar because it will keep moving energy on a subtle level and aid you in the transformation you desire.

Once you start working with crystals, you might be wondering if there are specific places in your home where you can place them for good feng shui energy. There are many spots in your home that can get better energy from specific crystals. Let me share with you the three most important areas.

1. Your front entrance. As a doorway between the outer and the inner world, the front door is considered very important in feng shui and needs strength, vitality, and protection in order to contribute to a good feng shui house. Placing crystals or stones with strong, protective energy in your main entry, as close to the front door as possible, will help ground and clear the incoming energy. As you know by now, the front door is also very important in smudging rituals. Examples of protective crystals include tiger's eye, black tourmaline, shungite, hematite, and smoky quartz.

2. Your bedroom. This often neglected room is one of the most important rooms for creating good feng shui energy in the whole house, as well as keeping your energy healthy and happy. In the bedroom, you want a combination of seemingly opposing energies. You need both fiery energy to fuel passion and nourish the senses, as well as peaceful energy to promote rest and rejuvenation. The absolute best crystal to have in the bedroom is the soothing rose quartz for its properties of heart healing and opening. You can also look into the warm and fiery carnelian stone for its ability to bring a calm sense of happiness and strengthen your physical energy.

3. Your living room. As one of the busiest rooms in the house, the living room can definitely benefit from the presence of crystals to both clear the area of potentially hectic energy, as well as bring more vitality and joy into it. Ammonites have an uncanny ability to constantly recirculate energy and create more harmony in the area where they are placed. An amethyst geode or cluster can bring cooling, calming, and inspiring energy. If you feel like your living room tends to get too busy and its energy chaotic and uninspiring, the beautiful celestite can help. It is a crystal with the ability to clear the energetic fields around it and reduce any chaotic emotions. Celestite can also improve communication, gift a person with sweet and calm confidence, and instill an unshakable belief in being loved.

How to cleanse and charge your crystals

It is important to cleanse your crystals regularly so that they can keep expressing their healing energies in your home. You can create a schedule for clearing them, such as once every month or two, and you can also just tune in to their energies and feel which ones are a bit overloaded and need cleansing. There are many ways to cleanse your crystals. All of them use the power of natural elements to cleanse and purify, as well as the power of seasonal cycles.

Here are several ways to cleanse your crystals. You can choose just one way, or you can alternate to experience which way feels best for you. I recommend ending the cleansing with a gentle smudging of your crystals, right before you place them back to their usual place in your home.

- Submerge your crystals into a bowl of pure water with some salt in it. Leave them there for a couple hours, then rinse under running water and let air dry.

- Rinse your crystals under running water then place under the sun, ideally in between 11 a.m. and 1 p.m. when the sun's power is the strongest. Notice this method does not work for all crystals, as sun will fade color in such crystals as amethyst and rose quartz.

- Rinse your crystals under running water, then place them under the moonlight from 11 p.m. to 1 a.m. You can also keep them there until morning, but be sure to bring them in before the sun rises.

My favorite way of cleaning crystals is to rinse them under running water and then place them overnight under the waning moon. I usually put them in the soil of a plant, and sometimes on some fresh flowers and leaves, as I feel this way they get rejuvenated with earthy energy. After all, this is where they come from, so it is like a sweet reunion! The only crystals I cleanse by using sunlight are my clear quartz crystals, and sometime a special rose quartz wand, but for a very short time. I never place my amethyst crystals in direct sunlight.

One of the quickest and most powerful ways to cleanse crystals is actually with sound. I mentioned the story of a crystal healer who cleanses all her crystals by placing them on the table and using the sound of singing bowls. I use the singing bowl cleansing method in combination with smudging and other methods. As you see, there are quite a few ways for you to choose from in deciding what feels best for you and your crystals.

As for the process of charging the crystals, this is actually quite easy. After cleansing specific crystals, hold them in your hands, close your eyes, and energize them with your intent. For example, if you have just cleansed your rose quartz crystals that serve as your feng shui cure for love—in other words, as your guardians and attractors of love energy—take a minute or two to hold them in your hand and imbue them with the energy you want them to hold. You can just feel the energy of love, visualize specific images that evoke this energy in you, or say words that evoke the energy of love. After you charge your crystals, place them back in the love and marriage area of your home or your bedroom to continue their work as feng shui cures.

Where to get smudging supplies

It is wonderful if you can buy all your supplies, especially crystals, in person, be it at a local bookstore, crystal shop, or a healing center. However, there might be cases when this is just not possible. Luckily for us, we live in a world where almost anything is available online. If you do not have access to a good local supplier for your smudging and space-clearing needs, look up the many online resources to help you.

Do your own due diligence and ask questions when you need to know more about the quality of your supplies, as well as where they are coming from. In time, you might decide to make your own smudge sticks and collect your own herbs, but to start with, find a good, trustworthy supplier, either online or in person.

I buy all my supplies from a wonderful local bookstore, as well as whenever I travel and find something new. Even though I bought most of my crystals in person, I have ordered some tumbled crystals online and was very happy with them. I have also placed many online orders for specific incense sticks and essential oils. I usually place my orders via amazon.com as they have a variety of retailers with really good choices.

Chapter 6

MAKING YOUR OWN SMUDGING SUPPLIES HAS POWER

Making your own smudge stick is a beautiful process. It adds more power to your energy, as well as to your smudging ritual. If you have any of the herbs available, I highly encourage you to create your own smudge stick. You can also make your own smudging blend and craft a very special smudging fan. I will show you how, step by step.

Start with your choice of herbs

The variety of herbs used for smudging is quite amazing. From sage to pine, juniper, and lavender, make your choice based on what is available in your area. Not all of the herbs listed here are used in smudging as practiced by Native American tribes. Some herbs, such as rosemary, vervain, and mugwort, for example, have a long history of use in European cultures. You can use just one herb to create your smudge stick, or combine several different herbs. Use your intuition to guide you. It might be helpful to know that in most Native American traditions, the four sacred medicines—white sage, cedar, tobacco, and sweetgrass—are never mixed. Out of the four sacred medicines, tobacco is the only one that is never used for smudging but either smoked during sacred ceremonies, or given as an offering.

Here are the most popular herbs used for smudging, along with their specific properties.

Cedar has always been considered a strong protective and cleansing medicine. It has been used for smudging and purification by many tribes. As one of the most ancient trees on earth, the energy of cedar is very potent, wise, and majestic. Cedar is considered a guardian spirit and a powerful healer, and it is often used in sweat lodge ceremonies.

Juniper is believed to be as ancient as cedar, with strong properties of protection and blessings. Juniper was also traditionally used to invite more abundance and prosperity. Smudging with juniper helps calm and purify the energy, as well as bring strong protective qualities to any space.

Lavender has a long use in smudging, too. Burning lavender sticks recreates a sense of cleanliness and a calm, deep joy. It brings a very sweet nourishing energy of blessings and being at ease with oneself. I always use it after I smudge with sage, as it feels like a sweet completion to smudging, like closing the circle in a sweet, unquestionable protection.

Pine has a grounding and purifying effect. It deepens one's breathing and cleanses one's mind. Using pine in smudging was considered to help bring forgiveness to any situation.

Sage is used for purifying, cleansing, and healing. It can bring mental clarity and is considered calming. Sage also brings the energy of blessings and protection. It is important to know that sage comes in many varieties, from black sage to purple sage to blue sage. There are over 300 varieties of sage, and not all of them are suited for smudging. White sage, also known as sacred sage or California sage, is the most popular variety of sage used for smudging. I love the scent of white sage the most.

Sweetgrass is a traditional smudging herb that has been used for centuries for its beneficial properties. The use of sweetgrass evokes a sense of trust, calm, and peace. Sometimes called vanilla grass because of its sweet, nourishing, and gentle energy, this herb is often associated with motherly energy. It has been called the hair of Mother Earth by Native people. Traditionally, sweetgrass is braided in three strands representing kindness, honesty, and love.

Rose petals are often used along with lavender in smudge sticks, so you can add a few petals to your smudge stick for beauty, as well as a special touch of graceful energy. Rose has the energy of harmony, love, and heart healing. It goes well with wild sage, rosemary, thyme, and lavender.

Rosemary was long used to protect from negativity, so it has powerful cleansing properties. Rosemary is also very invigorating and energizing. It is best to use rosemary combined with sage and other herbs in smudging. Traditionally, rosemary was considered a strong protector of the space, as well as capable of promoting the energy of devotion between partners.

Thyme is often used along with rosemary for purification and protection. Thyme is also used to bring more vitality, courage, and confidence, as well as to ease the energy of sadness.

Use your intuition to mix and match your herbs. Some herbs will go along just fine, and some will not feel very good together. Personally, I do not like bundles with too many herbs. I might admire their look—especially when they have flowers and colorful strings—but I feel it is good to give each herb some room to do its own healing job.

Craft your own smudge stick

The first step in crafting your own smudge stick is to decide on its look—its length and thickness, the specific herbs you will be combining, as well as the type of string you will be using to tie your bundle.

You can choose to make big smudge sticks about 2 inches thick, or to create thinner ones of less than 1 inch in diameter. Usually, thick bundles will smolder more slowly, which is helpful if you plan to smudge bigger areas, such as your whole home instead of just yourself or a small room. If this is your first time making smudge sticks, you can choose to make both thin and thick smudge sticks in one sitting. This way, you can experiment and see which thickness works best for your smudging rituals.

Along with various thicknesses of smudge sticks, there is also quite a wide variety of lengths. The shortest smudge sticks I have seen were about 3 inches and the longest about 12 inches. Again, choose what you feel will work best for you or create several stick in different lengths. Small smudge sticks are great for travel, clearing your own energy, and to give as gifts. The bigger smudge sticks are ideal for bigger jobs, such as a whole house smudging or clearing the land.

When choosing a string for your smudge stick, be mindful of the fact that the string will burn, along with the herbs, during smudging. This makes it very important to choose a string made from natural material such as cotton or hemp. You can choose strings in many bright colors, or go for a simple undyed string; it is all a matter of personal preference. The length of the string should be about 3 to 5 times the length of your smudge stick, because you will be tying it around your bundle several times.

While the thickness, the length, and the look of your smudge stick are up for play and experimentation, the non-negotiable part is that your smudge stick has to be wrapped tightly. The herbs will dry and shrink a bit in size, so be sure to do a good job, as you do not want your stick to fall apart while you are smudging. You also have to be sure that the string is, as mentioned before, made of natural material. The last thing you want is to pollute the air by burning synthetic strings during smudging! Here are the supplies you need in order to make a smudge stick:

- 🌷 The herbs of your choice

- 🌷 A natural string, either colored or plain

- 🌷 A pair of scissors

Once you have your supplies, follow these simple steps.

1. Place your supplies—the herbs, the string, and the scissors—on a flat surface. Arrange your herbs into several bunches of chosen length and thickness. If you are combining different herbs, such as sage, rosemary, lavender, and rose, be sure to arrange them in a visually pleasing way. Usually, you will place the rose petals or the lavender flowers on the very top of your bundle so as to create a really attractive smudge stick.

2. Prepare the strings by cutting them into the right length, which is 3 to 5 times the chosen length of your bundles. Be sure to have enough strings for the number of your sticks.

3. Tie a knot around the stems of your herbs to securely keep them in place and then wrap your chosen string around the stems several more times. Holding the stems with one hand, use your

dominant hand to tie the string on an angle from the base of the stems to the tip of your herb bundle. Then, reverse the angle and tie the string back to the base of the stems. This way you will create the crisscross pattern seen in many smudge sticks. You can choose to go up and down several times, or just do it once; at this point, it is all a matter of visual preference, as long as the herbs are secured tightly! When finished, tie another knot to secure the wrapping. To make the process more powerful, you can choose to make nine loops as you string upward toward the tips, and nine loops as you tie downward toward the stems. You can also choose to wrap the base nine times to finish.

4. Place your sticks to dry in the shade for about a week or so. There are two basic ways to dry your smudge sticks and two important rules to follow. You can place your smudge sticks on a drying rack or screen with plenty of air circulation, or you can hang the bundles from a cooking rack, a wall handle, or any other makeshift creation you can come up with to help hang your smudge stick securely. The two important rules to follow are to be sure your dry them away from the sun, so the area you choose has to be reasonably dark, and there has to be plenty of air circulation.

Your precious smudge sticks will be ready for use in 7 to 10 days.

How to make a smudging fan

Making your own smudging fan is not difficult. You can use your creativity and craft an elaborate smudging fan with many feathers and decorative elements, or make a simple fan with a few accents. The sky is truly the limit when it comes to designing your smudging fan. Some of the most common items used in the adornment of a smudging fan are leather pouches and cords, beads, crystals, twigs, and shells. Any treasure from nature that speaks to your heart—and is resistant enough to adorn the fan—will work. I have seen ocean-themed smudging fans with a variety of shells and dried sea vegetation, forest-themed ones with twigs, antlers, and small pinecones, as well as fiery smudging fans with bright red-colored feathers, golden tassels, and purple crystals.

You can also just use one feather as a smudging fan and embellish its base with a leather cord and small beads or crystals. Using a feather or a fan instead of using your hand to wave the smoke does make a big difference during smudging. It brings a light and soaring energy that feels happy, inspiring, and comforting, too. I feel like my personal energy is being gently brushed, cleared, and uplifted when I use my smudging fan.

Made from bird feathers, a smudging fan brings the uplifting energy of birds to any smudging ceremony. The symbol of birds is revered in all cultures as messengers from the Spirit world and the creatures closest to Heaven. Soaring freely in the sky, the birds are a call to remember your Soul's freedom. "Soar like an eagle" is a beautiful blessing to say to someone while smudging them. Traditionally, the smudging fans were created from the whole wing of the eagle, so the idea of a smudging fan is to replicate the wing of the bird.

If you are want to make your own smudging fan, here are the supplies you'll need:

- A variety of feathers, either collected from your walks in nature or purchased from a craft store
- A small and sturdy branch, driftwood, or a piece of wood to be used as a handle, typically between 2 to 5 inches long
- A glue gun
- A cord made from leather or another material that you like
- Embellishments such as crystals, small beads, personal talismans, or any other items that appeal to you

When you have all your supplies ready, follow these simple steps:

1. Decide on the surface on which you will create your smudging fan. A large kitchen or craft table is ideal for this purpose. As you will be using glue, be mindful to protect the table surface, if needed.

2. Visually divide your surface in four areas and place all your feathers in one area, all your embellishments in the other, the branch to be used as a base and the glue gun in the third area, and leave the fourth area open to be your actual creation area.

3. Pick the feathers you would like to incorporate into the design for your fan and place them in the open area of your table. Some of the feathers will be longer than others, so play until you find their best arrangement. Be sure they are all facing the same direction, resembling the wing of a bird.

4. Once you have arranged the feathers in the desired design, pick up your branch that will be used as the base of the fan. Decide which side of the stick will be used as the front of the smudging fan. Pick your feathers, one by one, and glue their quills to the top of the branch, according to your vision. A glue gun is most helpful here. Be sure to hold each feather in its chosen place for at least 20 seconds before the glue dries.

5. Once all feathers are firmly placed on the branch in the way that is most visually pleasing to you, pick up your leather cord. Glue one end of the cord to the top of the stick (where all feathers are gathered) and tightly wrap the cord all around the stick. You can choose to cover the branch completely or only partially; the design is completely up to you. Use glue to hold the end of the leather cord firmly in place.

6. Choose the embellishments that will look best on your smudging fan, be it a small tumbled crystal, a seashell, a brightly colored small feather, a string of beads, or any other items that speak to your heart. Place the items on the base of your smudging fan to find their best positioning. Once you are happy with the look, use the glue gun to secure them firmly in place.

Your smudging fan is ready!

I recommend that you smudge your fan gently and leave it for a bit on your altar before using it. The best place to display your smudging fan is on your space-clearing altar. If you are traveling with your smudging fan or just need to store it, be sure to find a proper way to pack it.

Making smudging blends

A smudging blend is really easy to make. It is similar to making a tea blend: All you have to do is mix your choice of dried herbs and place them in a dry container away from heat. You might want to experiment by adding or omitting herbs from your blends. When not sure, it is always good to start simply and just mix a few herbs such as sage and lavender, for example.

My smudging blends are very simple. I mix dried white sage, rosemary, lavender, and rose. Sometime I might add cedar or pine, although most of the time I use them separately. I like to display a handful of my smudging blend in the abalone shell on my altar.

Chapter 7

SMUDGING ALTERNATIVES YOU CAN USE

If you are not in love with the scent of sage or several other herbs typically used in smudge sticks, there are many alternatives to smudging. There are quite a few options you can still use to clear the energy in your home. I love the scent of sage, both white sage and wild sage; it centers me and gives me a deep sense of calm and groundedness, but I know some people do not like this scent. Respect your preferences and do not force yourself to smudge with sage if you honestly do not enjoy it. There are other alternatives that I will share with you here, such as working with palo santo, a sacred wood from South America; burning resins; diffusing pure essential oils; and more.

Even though sage is the most popular herb used in smudging, not all smudge sticks have to have sage and traditionally, not all Native American tribes use sage in their ceremonies. A variety of herbs were used depending on what was available in the area. From sweetgrass, tobacco, cypress, and juniper to piñon and pine needles, there are many choices when it comes to smudging your house with herbs.

In addition, because the ritual of smudging is used all over the world, we also have access to smudging alternatives from cultures far and wide. Let us explore some of the most popular alternatives to traditional smudging, the ancient ways to cleanse, purify, and ward off negative energy, as well as attract healing energy to your home.

Pick and choose what you love the most.

Palo santo, the sacred wood with blissful aroma

I fell in love with the aroma of palo santo from the moment I first inhaled its scent. This was quite a few years ago while traveling in Ecuador with my good friends. The scent was heavenly for me—a rich combination of the warmth of the fireplace, the sweetness of sun-drenched summer scents of berries and flowers, with the feeling of a glorious sunset. I was mesmerized and needed to know more about these mysterious wood chips used for smudging. I managed to bring home several palo santo sticks and treasured them for quite some time until they appeared for sale in North America, too. Now you can easily buy them online and in most new age stores. They are very affordable, last a long time, and bring powerful, healing energy to your home. They are also very easy to use in your smudging and space-clearing rituals.

The story of palo santo, or Holy Wood, is quite powerful. To start with, these healing sticks used for smudging come only from fallen palo santo trees or branches. I was told that for the stick to be truly healing, the tree or the fallen branches that it comes from have to lie on the forest floor for up to 10 years. This activates the medicinal and mystical properties of this powerful tree.

Palo santo has been used by South American shamans for centuries for the purpose of healing, as well as for visioning during sacred medicine journeys. There is a sense of maturity and visionary power combined with the energy of sweet protection that comes from smudging with palo santo. It brings peace and clarity to your mind and helps connect and receive guidance from the Spirit world.

The tree itself belongs to the same family as frankincense and myrrh and is revered as an ancient and potent medicine. While the energy of palo santo is very different from sage, the effects achieved by smudging with palo santo can be similar and very powerful; I highly encourage you to use it in your smudging sessions. The cleansing power of palo santo wood comes with a sense of sweet blessings and nourishment; it clears the negative vibrations and brings forth a healing quality of energy to your home.

In addition to the use of palo santo sticks for smudging, the tree also produces a very aromatic and potent essential oil with many therapeutic uses, from strengthening the immune system to relieving anxiety and depression. I have also seen palo santo incense on the market recently, as well as small shavings of the wood to use for burning, so there are several options for you to explore. My favorite one is to use the palo santo wood sticks.

It is very easy to use these aromatic sticks. Similar to an incense or a smudge stick, you light one end of your palo santo stick from a candle. Let it ignite and burn for less than a minute, then gently blow out the flame. Holding the stick in your dominant hand, walk around your space waving the sweet smoke into the air. You can smudge your own energy with palo santo the same way as you would with a sage smudge stick. One palo santo stick goes a long way as, unlike incense, you do not use the whole stick during smudging.

No matter if you buy palo santo in person or online, I highly encourage you to buy from a reputable supplier with proven ethical practices of harvesting. There are two ways these powerful sticks find their way to you: They are either harvested in the wild or come from a tree farm. While several countries, such as Ecuador and Peru, have implemented mandatory laws against harvesting live palo santo trees, their newly found popularity in the West has created illegal cutting of live palo santo trees in the Amazonian forest for profit. Be mindful of that and inquire how your palo santo sticks were harvested. If your palo santo is wild harvested, it has to be done responsibly and respectfully. If it comes from a palo santo farm, chances are they have responsible practices.

Burning sacred resin in your home is powerful

The ritual of burning sacred resin is very old, powerful, and fascinating. We all know the story of frankincense and myrrh being brought to baby Jesus. Their power was considered far more valuable than gold.

Sacred resins are resins from healing trees, the most ancient ones in use being frankincense, copal, and myrrh. Each of these sacred resins has its own unique properties, yet at the same time, they create somewhat similar effects. Frankincense is considered cleansing, protecting, and uplifting; myrrh has the ability to clear confusion and align a person to the sweetness of their heart; and copal clears the mind and removes energetic blockages to attract positive changes into one's life.

All three of these sacred resins are used for spiritual cleansing and protection. They are also used for opening the way for beneficial energy, whether it is the energy to attract more love, inner peace, or better health. Often these resins are mixed together for a more potent clearing, and sometimes other herbs are added to the mix as well. As with all herbs, woods, and resins used for smudging, your intent is important while burning sacred resin because beneficial energy comes in so many forms!

While resin is a good alternative if you do not like the scent of sage and other herbs traditionally used in smudging, burning resins requires a bit more work. It also takes a bit more time and preparation, as well as more supplies.

If you have never burned sacred resin before, it is worth trying, as the scent is amazing. It connects you to ancient roots and mystical memories that have no words to them; just strong feelings. I find the ritual of burning resin very centering and grounding, and at the same time deeply mystical. Here are the supplies you need for burning sacred resins:

- A fireproof, heat-resistant bowl. You can also choose to go for an elaborate brass incense burner called a censer that is usually adorned with intricate designs. You might have seen them in many new age shops, some of them resembling the look of Aladdin's magic lamp. Censers come in all shapes and sizes, and you can choose between designs with ventilated lids or without lids altogether. However, there is no need to go for intricate censers in order to burn sacred resins. To start with, a good fireproof bowl is all you need. I use a small marble bowl.

- A bit of sand to place in the bowl. The sand helps absorb the heat and protects the surface of the bowl from overheating or burning.

- Incense charcoal. This is the heat source to burn your resins. You can buy charcoal pieces online or in stores that sell incense and smudging supplies. They are about 3 to 5 dollars and come in a package of 7 to 10 pieces.

- A pair of tongs. Using tongs is a smart way to go when burning resin so as to avoid burns. You might have to hold the charcoal in the flame for more than a few seconds, so never hold the charcoal in your hand when trying to light it; always use tongs.

- Your choice of sacred resin. It may be frankincense, myrrh, copal, or a mixture of the three. You can also add some loose herbs if you so desire.

- A candle.

Here are the six steps for burning sacred resin:

1. Place all your supplies within reach and on an even surface.

2. Fill your fireproof, heat-resistant container with sand.

3. Light the candle. Then, carefully pick the charcoal with the tongs and hold it over the candle flame in order to light it. You might have to hold the charcoal in the flame for a bit before it actually lights up.

4. Once the charcoal is lit, place it on top of the sand. If it does not light evenly, you can help it by adding flame to its edges.

5. Wait a minute or so for the charcoal to form a layer of gray ash. This is a sign for you that the charcoal is ready to burn resin.

6. Now comes the fun part! Place a small amount of your sacred resin on top of the charcoal and enjoy its healing fragrance. You can keep adding new resin after the old layer has fully released its power.

If you are just starting with burning sacred resin, it is best not to move your container; just have it placed firmly on your chosen surface. Once you become more confident with this ritual, you might choose to move it from room to room. You can do that by placing your fireproof container on a different surface, such a small tray, to give extra protection to your hands while moving it.

Please be mindful to not use abalone shells for burning resin, as the shell is not fireproof and it can get damaged. Censers can be excellent for this purpose. They can also be a good addition to your space-clearing collection if you fall in love with burning sacred resins. There is a wide variation of censers, from the big ones you see in Catholic churches and Taoist and Buddhist temples to the smaller, ornate, and creative censers used daily in many homes around the world. As with smudging, you can buy all these supplies either online or at a good quality bookshop or new age store.

It might take you a bit of experimentation with the process of burning resin. Sometimes your charcoal will not light up easily, or once lit, it might not maintain its heat. The good news is that even though the process sounds complicated, you can easily master it, and then enjoy the absolutely divine aroma of sacred resins in your own home.

Liquid smudging is easy and convenient

Liquid smudging is the easiest way to perform smudging without actually having to smudge! While the effects of liquid smudging are not the same as traditional smudging because the powerful energy of the Fire element is missing, you are still creating a considerable shift in your space if you decide to choose liquid smudging.

The term "liquid smudging" means using a liquid made by mixing one or several essential oils in pure water. The most popular liquid smudging options are ceremonial white sage spray and palo santo spray. You can also find mixes that contain a variety of essential oils, as there are many potent oils that can clear the energy in your home. All you have to do is buy them or make them yourself, it is that simple.

For example, the one I am using now has eleven essential oils in it, including cypress, rosemary, vetiver, devil's club, and patchouli. Another one that I also love has just the white sage essential oil in it. There are many options you can look at if you decide to go for the option of liquid smudging either instead of, or in addition to, traditional smudging.

Liquid smudging is slowly growing in popularity, so there will definitely be more and more choices available on the market. The biggest benefits of liquid smudging are its practicality and portability—you can take it with you anywhere and do it really quickly, be it in your hotel room, your car, or your place of work.

You can also decide to make your own liquid smudge. Choose your own selection of essential oils, or follow the recipe below, which has white sage, atlas cedar, and lavender essential oils.

To make your own liquid smudging blend, you will need a 4-ounce bottle with a spray cap, and preferably dark-colored glass. These can be bought at specialty essential oils or body care stores, or you can reuse your facial toner bottle. If your bottle is larger, make sure that you increase the proportions of this recipe accordingly with its size. For instance, an 8-ounce bottle would require double the amount of each ingredient in this recipe. Here are the supplies you need:

- 4-ounce bottle
- 25 drops of white sage essential oil
- 15 drops cedar essential oil
- 15 drops lavender essential oil
- 4 ounces pure water

1. Ensure that your bottle is thoroughly clean.

2. Add the essential oils to your bottle. Add water.

3. Screw on the spray cap. Shake the bottle thoroughly.

Your liquid smudge is ready! Remember to shake the bottle before each use.

You can play and experiment with the selection of oils to add to the main oil, which in this recipe is the white sage essential oil. It is best to avoid adding too many oils in order to help keep the main note, the voice of the white sage, really strong. Other essential oils that go well with sage and can amplify its cleansing effects are citrus oils such as lemon, wild orange, and lime. The essential oils of frankincense, cedarwood, geranium, and sandalwood also go well with the white sage.

Adding a bit of unprocessed sea salt to your liquid smudging mixture can amplify its effect in space clearing. You can also add a small crystal of your choice to your bottle. The best way to do that is to place the crystal in a bowl of water and let it sit in the sunshine for at least three days. Add this charged water, along with the crystal, to your liquid smudging spray.

Using solar-charged water and crystals brings the benefits of the fiery energy of the sun to your mixture. Of course, the crystal has to be tiny enough to fit in the bottle. You can also use the solar- and crystal-charged water without actually placing the crystal into the bottle. Water has memory and it will keep carrying the vibrations of both the sun and the crystal, so see what feels and works best for you.

Another alternative is to use *Agua de Florida* (Florida water), which is a 19th-century formula that blends various floral essences and aromas in an alcohol base. This blend has become quite popular in recent years and can be found either online or in many new age stores that sell smudging supplies. Many shamans use *Agua de Florida* for cleansing and protection, as well as for grounding their energy after deep medicine journeys. I love the scent and energy of this blend and use it often to quickly refresh my energy. Basically, I just rub a couple of drops into my palms and then wave the aroma around my heart and my head. *Agua de Florida* can also be used to refresh the energy in any given space by mixing several drops of this blend in a water bottle with a spray cap and then spraying the aroma into the air.

Even if you love traditional smudging and enjoy it regularly, a liquid smudging mix can help you while on the move, whether during travels or when quickly needing to shift the energy in your home or office. I use all the alternatives that I share with you here and I also love traditional smudging. It does not have to be an either-or choice, as it is a matter of finding what you love the most, and what works best for you and your home at any given moment in time.

Consider natural incense

The art of making incense is ancient and there are complex recipes and formulas that are followed in making powerful incense. From ancient Egypt and India to Tibet and Bali, from Nepal to Japan, the use of incense is a very important part of many spiritual rituals.

If you have ever had a bad experience with the scent of an incense, I encourage you to give the idea another try. Since the use of incense is very popular, there is an overwhelming amount of low-quality incense on the market. Incense can be a potent recipe made of naturally healing ingredients, or it can be made from a quick formula of several artificial ingredients. It is good to be mindful of that when you are choosing incense for your home. There are quite a few forms of incense available on the market—from the most popular thin sticks that come in various lengths to a variety of small cones, rectangular bricks, incense coils, and even mysterious-looking small twisted ropes, a popular Himalayan incense.

There is an even wider variety of incense holders you can buy that come in all shapes, sizes, colors, and materials. I have seen beautiful incense holders in the form of praying hands; various animals such as birds, cats, and mythical dragons; and even Buddha head incense holders. (I would never want to burn incense on the Buddha's head, though!)

After experimenting with many designs, I have found what works best for me and what I can highly recommend. I love the simplicity of rectangular incense holders and was very happy when really wide ones came on the market. If you are used to burning incense, you know how frustrating the use of thin incense holders can be: After your incense stick is done burning, you can find ashes everywhere around the holder rather than just in it.

Along with wide incense holders—mine are about 5 to 6 inches wide, as compared to the typical 2-inch incense holders—there is another way to burn incense that looks beautiful and works well, too. I love filling a small bowl with sand, small rocks, or with various grains such as green or yellow peas or green dahl beans, and placing an incense stick or cone in the center of the bowl. There is something powerful and earthy about having a source of food as a base for burning incense—the incense becomes an offering and brings a sense of humility and gratitude to the whole process. After the incense is done burning, I mix its ashes with the grains and place a new incense stick or cone in the center. The beauty of this method is that you can use several incense sticks at once for stronger purifying effects.

When it comes to choosing the best incense, it is always best to buy it in person. You will not only get to ask the seller for their recommendations, you can also have a direct experience

with the variety of natural incense scents. Afterward, if you like your natural incense, you can order it online if visiting the store is not convenient for you. I always like to try new incense when I visit my favorite local shop, even though I am also very faithful to the ones I love and have been using for a long time, called the Nag Champa and the Woods incense.

In addition to natural incense made of various blends such as oils, herbs, and resins, you can also find amazing incense made of just one herb or wood or resin. I have been using cedar and balsam fir natural incense cones for a long time in our home. Their scent feels especially soothing and quite magical on long winter nights; it is cleansing and energizing at the same time. I have recently added to our collection tiny bricks of incense

made from piñon, juniper, hickory, mesquite, and alder, each brick having a single scent. It is quite a sensory experience! I am very happy to see white sage incense appearing in the market, as well as palo santo incense. This makes the job of clearing the energy in your space really easy and pleasurable.

To use incense as an alternative to smudging, all you need is your choice of natural incense, a good incense holder, and a lighter. Light the incense, let the fire burn for a bit, then blow on it to extinguish the fire and let the incense smolder. Place the incense stick in the holder on your altar or any other space you want to cleanse and voila—this is all the work you have to do!

If you need deeper clearing, using incense will not be as powerful as smudging. However, if you are pressed for time and need stronger clearing, here's what you can do. Hold several high-quality incense sticks in your hand like you would a smudge stick and light them from the candle flame. After extinguishing the fire to let the incense smolder, hold the sticks in your dominant hand and move from room to room or around the area you want to cleanse. Spread the smoke the same way as you would when smudging with the smudge stick. This will create a stronger smoke than just one incense stick would and will help cleanse the energy. The key here is to choose natural incense, so potentially, a high-quality white sage or palo santo incense should help you achieve the same result as smudging.

I remember many years ago while preparing to teach a feng shui class and doing my typical clearing of the energy in the room, a lady came in early that made me feel uneasy. She looked nervous and agitated, and her behavior was quite unfriendly. I decided to focus my attention in the positive direction, as I knew that she would either change her energy during the class, express what bothered her, or leave the class—none of which was in my control. At the end of the class, she raised her hand to share her experience. It turned out that while seeing the incense stick I had burning in one specific corner of the room, she felt an almost irrational fear. Her experience with incense was awful because it always gave her strong headaches. However, this time was very different, she said. Not only was she very happy and pleased that she did not get a headache by staying in the room, she also felt energized and happy, and had a clear mind after the class!

I am sharing this story to illustrate the power of natural incense and to encourage you to try it even if you have had unpleasant experiences in the past. A natural, high-quality incense made by a reputable supplier will not only clear the energy in the space, but also evoke a sense of freshness and sweet calm.

This story also goes to illustrate the power of space clearing, which I always do for any space I am using for teaching. Clearing the stagnant or negative energy accumulated in a space and inviting fresh, clear, and vibrant energy is something you can easily do, too. In this specific case, I did not use smudging, as it was not appropriate for the space (a classroom on a university campus). I only used incense, liquid smudge, as well as the power of sound (clapping and high vibration music), which I shared with you in the previous chapters.

In addition to the ease of using incense, I also love to sometimes just sit quietly and watch the incense smoke waver slowly in the space. Since you do not have to do any work other than place the incense in the holder and light it, this can be a space for contemplation and a meaningful repose. There is something very soothing, calming, and even magical in quietly sitting in front of the burning incense and watching it all unfold. The silent dance of wave after wave of smoke gracefully flowing into the air, the warm amber glow of the incense stick, and then the still quietness of the end, the moment of emptiness—no more smoke, no burning incense, just the memory of experience. It makes you more present somehow, more aware of the uniqueness of each moment, more grateful for every breath. This might be the reason why incense is continuously used in many temples around the world—I find its unique energy encourages mindfulness, peace, and gratitude.

You can always use essential oils

The use of essential oils is an all-time favorite of mine, especially since the creation of ultrasonic diffusers. I cannot even recount how many beautiful candle oil diffusers I have burnt because I got busy and forgot to refill the tiny dish! If you do not have an ultrasonic aromatherapy diffuser in your home, I highly encourage you to get one. There is such a wide variety of diffusers on the market today. Their prices are getting lower, while the designs are getting better and more sophisticated. The beauty of an ultrasonic diffuser is that it uses cold mist to diffuse essential oils into the air, so the scent is more potent as compared to candle diffusers. The best part, however, is that it does its brilliant work without any attention from you whatsoever! All you have to do is fill the container with pure water, add your choice of essential oils, and choose the function or the timing. It is the easiest and most powerful way to continuously diffuse essential oils into the air in order to purify and energize it. The effect might not be as powerful as it would be when smudging, burning sacred resin, or burning incense because the fire element is missing. However, the clearing will still happen, even if on a more subtle level.

Here are some of the most popular essential oils used for clearing energy. You can use them as a singular note, or combine them with other oils as your intuition guides you. I have also included some suggestions for specific combinations if the idea of creating your own mix of essential oils feels intimidating.

Balsam fir oil has many uses—from fighting infection and relieving muscle aches to uplifting and purifying the energy in any given space. Balsam fir is refreshing, comforting, and balancing. It helps one deepen the breath and connect to the source of inner wisdom.

Cedarwood oil is soothing, warming, and grounding. Its woody scent can be used to clear the energy in any space and imbue it with peace and protection. Cedarwood brings calm and strength, and is considered an ancient symbol of wisdom and abundance.

Cypress oil helps remove anxiety and stress and clears the way for holistic mind-body healing. Its clean and energizing aroma can be used to clear the energy in any space and imbue it with vitality and calm.

Eucalyptus oil is famous for its multitasking talent, as it is antibacterial, antimicrobial, antiviral, antifungal, and anti-inflammatory. It is one of the most popular essential oils because it can help in so many ways—it can rejuvenate one's energy, improve memory, reduce tension, and clear the air, of course! I often use eucalyptus oil in my diffuser when I have a lot of writing to do. It keeps me alert, energized, and inhaling deeply.

Frankincense oil comes from the wild-crafted resin of *Boswellia carteri* trees. It has been used for thousands of years for its beautifying and purifying properties. Frankincense feels both comforting and energizing, it raises the vibration in any space, and it promotes a deep sense of well-being. With regular use, frankincense is considered to improve memory, reduce inflammation, and promote healthy sleep.

Juniper berry oil is known for its detoxifying and immune-boosting properties. Along with its ability to remove negativity and purify the air, juniper is also very relaxing and can help promote good sleep. Juniper oil helps clear negative energy and creates a sense of peace and protection.

Lavender essential oil is definitely the most popular, as well as one of the most versatile oils. Its cleansing, purifying, calming, and soothing properties go along with its ability to revitalize and energize. Lavender is an adaptogenic herb, meaning it will work with your energy level to provide what is most needed in order to bring it back to balance—either activate it or calm it down. If you are new to the use of essential oils, lavender is definitely the one to start with.

Palo santo oil is both uplifting and grounding. It has the ability to cleanse and purify by transmuting low and negative energies in a space or in someone's energy field. Palo santo also brings a sweet sense of peace and calm. It is used in ceremonies to help participants go into deeper states of meditation and communion with the Spirit.

Peppermint can quickly clear the tension in your body and in your environment. It cleanses and refreshes the air, clears the mind, energizes the space, and brings feelings of renewal and hope. It is one of the best add-ons to any essential oil mixtures you will make, as it tends to blend well and compliment most air-purifying oils.

Pine oil has a fresh and empowering scent with calming and uplifting energy. It has anti-inflammatory properties, helps relieve headache, and clears the air of pathogens. The use of pine oil can help transmute negative energy and bring a sense of renewal and hope.

Rosemary oil is revitalizing, purifying, and energizing. It helps release stress, promote mental clarity, and boost the immune system. Rosemary belongs to the same family as lavender and sage, which explains their similarity in clearing negative energy and creating a powerful sense of peace.

White sage oil is revitalizing, purifying, and cleansing. It has a long history of use in clearing negative energy. The earthy aroma of white sage calms the mind and can alleviate fears and anxiety. White sage is also antibacterial and helps fight off infections.

In addition to these oils, you can use essential oils derived from citrus fruits, such as lime, lemon, grapefruit, sweet orange, and mandarin, as they are also energetically clearing.

Let me share with you several essential oil blends to use in your aromatherapy diffuser. You can also use these mixes for your liquid smudging, if you desire. All you will have to do is mix the essential oil blend with water and use your spray bottle to "liquid smudge" any space. I use both liquid smudging (air mists) for a quick shift of energy in my space, as well as diffuser blends for a more extended effect.

There are only two supplies you need for your aromatherapy diffuser oil blend: a bottle, preferably with dark-colored glass, in which to store your blend, and your choice of essential oils. To make the blend, mix the oils in the bottle, close it tightly, shake it, and store it away from heat and light. This is especially important if you chose a bottle made of clear glass.

Here are two easy recipes for clearing negative energy with essential oils: a heavy clearing blend, and a lighter one for cleansing and attracting good energy. If these combinations feel a bit too intense, you can start by using just one or two essential oils in your diffuser to get a sense of what you like the most. Start by using 15 drops of your chosen blend in a 50 ml diffuser bowl. If you want a stronger aroma, increase the number of drops to 20 or 25. If your diffuser has a larger bowl, adjust the number of drops accordingly. The bigger the size of the bowl, the longer it lasts and the larger the area it can cleanse. Some diffusers have 100 ml bowls, and some have small 20 ml bowls, so adjust the ratio accordingly.

Heavy clearing essential oil blend

- 20 drops lavender essential oil
- 20 drops rosemary essential oil
- 15 drops lemon essential oil
- 10 drops peppermint essential oil
- 5 drops eucalyptus essential oil
- 5 drops pine or juniper essential oil
- 5 drops rose geranium essential oil (optional)

Cleansing and attracting good energy blend

- 30 drops tangerine essential oil
- 25 drops lemon essential oil
- 15 drops lavender essential oil
- 15 drops grapefruit essential oil
- 10 drops peppermint essential oil (optional)

If counting drops for your blend makes you nervous, know that most of the time people err on the side of adding more rather than less. You cannot really go wrong with it, so relax and allow yourself to enjoy all the work the diffuser does for you! If you feel like the scent is not strong enough, stop the diffuser and add a couple more drops of essential oil. It is that simple!

One note on the use of aromatherapy diffusers—even though they are very easy to use, diffusers still need some maintenance, especially if you use a variety of essential oils in them. From time to time, it is good to use rubbing alcohol to clean the interior of your diffuser's bowl. I know some people even let the rubbing alcohol go one full cycle in their diffuser by adding several drops of alcohol instead of essential oils in order to clean it. This does not resonate with me, so I do not do it. I just use a cotton ball with rubbing alcohol to cleanse the bowl of my diffuser, then let it air dry.

Unlike all the other smudging options, you do not have to move the aromatherapy diffuser from room to room. It has to be stationary to cleanse the air well. Since diffusers are quite affordable, you can have a few of them around the house; the bedroom, the bathroom, and the living room being the main areas to benefit from a diffuser. If you work from home, a diffuser in your home office can do wonders, too.

As for where to buy your aromatherapy diffuser, there are plenty of available options. You can start with an online search in order to check the prices and looks of various ultrasonic diffusers. Many stores carry essential oil diffusers, from health food stores to yoga studios, so you can experience how a diffuser works before you buy it.

Chapter 8

SMUDGING RITUALS FOR LIFE, FROM BIRTH TO DEATH

Even though the smudging process in itself is simple and always has the same structure, there are many additional elements you can bring into it to make it more powerful. You can also make smudging fit any specific occasion—be it the birth of a baby, the promotion you dreamed of, or when you are in need of strengthening your health and peace of mind. In many of the above cases, you might not be able to choose the best timing according to seasonal or moon cycles. However, I am still giving you suggestions for the best timing of each of these rituals, assuming your situation allows you more leeway with the timing of a specific ritual.

As you will see, you do not always have to smudge your whole house for these rituals, just specific areas. However, I hope that by the time you decide to use smudging for specific rituals, your house will already be benefiting from regular smudging sessions.

When you leave your old house

Leaving an old house can be a joyful occasion for many. It is our tendency to look forward to something new, fresh, unknown, and exciting, which a new house can definitely represent. Leaving an old house can also be a sad experience, whether you have to leave a house because your relationship has ended, or when you need to move to a new city even though you love where you live. While the emotions will be different in different situations, for the purpose of a smudging ritual, we are focusing on one factor that will be the same in all cases: the need to complete your energetic interaction with the house. Your relationship with your house is similar to any other important relationship in your life. If you leave a house, especially a house where you lived for a while, without a heartfelt and deep clearing ritual, your energy can feel incomplete on many levels.

You can do this ritual alone or with your whole family. If you feel like it will be more powerful for you to do this ritual alone so as to connect more deeply to your house, please trust your feeling. Each family member can find their own way of saying goodbye to the house, but there has to be at least one person who performs this deeply healing and necessary ritual.

Time it takes: 15 minutes to 1 hour

Best time of the day: Between 11 a.m. and 1 p.m.

Best time period: Waning moon

Best day of the week: Saturday or Sunday

Additional supplies needed: White- or pink-colored tall candle, five crystals

Set your smudging supplies on your altar. For this specific ritual, choose a tall candle that can last for many hours. It can be either a pink or a white candle. Include five crystals that have been recently cleansed; rose quartz and clear quartz are highly recommended. Clearly formulate your intent before you start smudging. In the case of leaving an old house, your intent is twofold—to express the energy of gratitude for everything you experienced in the house, and to ask for release and take with you the energy of the most beautiful, cherished moments experienced there.

Clearly express your intent before you start the smudging ritual, either silently or out loud. It can be a simple sentence, such as "I am so grateful to you for giving us warmth and shelter through all these years. Please release all the joyful beautiful memories that are contained in your walls, let me take them with us to our new home. I will always keep you in my heart." Or, maybe you feel the need to write your own prayer or even a long letter to declare the most cherished moments. Go with the expression that feels best to you. The most important thing is to clearly keep in mind your intent for this ritual.

Next, follow these steps:

1. Light a candle and place your crystals around the candle.

2. Silently ask the energy of all the good memories experienced in this house to enter the crystals and the candle flame.

3. Take a few deep breaths, then light the smudge stick from the candle.

4. Slowly smudge your own energy.

5. Go to the center of your house, also called the heart of the home in feng shui, and slowly face each of the cardinal directions, moving clockwise.

6. Smudge each direction and ask the guardians to help release your energy from the house in gratitude, so that you can fully move on into your new home. I suggest that you use the seven directions mentioned earlier: the four cardinal directions, plus Sky, Earth, and Heart by honoring them in gently waving the smoke in each direction. Then, spend a bit of time connecting from your heart to each and every area of your home.

7. If a specific area of your house calls you, start there. Often, the house will need you to spend more time in an area that has a lot of energy accumulated in it. If this is the case, you will feel a clear sense to go there. If not, go clockwise from where you are to smudge your whole house.

8. Keep expressing your love and gratitude while smudging each and every wall, room, and corner of your house. Also, clearly keep your intent of taking the best memories with you. Feel or imagine the energy of these memories gathering in the palms of your hands.

9. Come back to the center of the house and extinguish your smudge stick.

10. Place your palms above the crystals and the candle. Visualize the loving, beneficial energy you just collected in your hands entering both the candle flame and the crystals. Amplify it with golden energy from your heart.

11. When ready, extinguish the flame, and spend some time in silence evaluating your experience. If you need to write down specific insights or memories, now is the time to do that.

12. Mindfully pack your candle and your crystals, as you will need them for the ritual of moving into a new house.

As with any ritual, be sure you have some time alone after you complete it. This is necessary in order for you and the house to integrate the experience. You have just created a huge shift with your intent, so it is important to be gentle and allow the Spirit to create new pathways in your life.

Moving into a new house

Most of the time, the occasion of moving into a new house is happy, energizing, and full of the energy of fresh, new beginnings. It is a possibility to start anew, to change some old habits, and to feel more optimistic about life in general. In many cases, we have also cleared a lot of clutter before we left the old house, so everything is lighter and more flowing.

As with any new relationship, it is important to start it mindfully and with a clear intent. If you have done the "leaving your old house" smudging ritual, you will have your candle and crystals to provide an energetic foundation for your new home. They will also provide a sense of continuity and create a safe container for all beneficial energies from your previous house to be welcomed into the new one.

If for any reason you could not do the leaving your old house ritual, start with a new candle, and use powerfully charged crystals. Charge your crystals with both the power of natural elements, as well as with the power of your intent.

Time it takes: 15 to 30 minutes

Best time of the day: Between 7 a.m. and 9 a.m.

Best time period: Waxing moon

Best day of the week: Wednesday or Thursday

Additional supplies needed: Singing bowl or bell, fresh flowers, a bowl of water or an essential oil mix with citrus oils

1. Place your smudging supplies, as well as the candle and the crystals, on your altar, ideally in the center of your new home. You might also want to use a singing bowl or a bell, as well as bring the presence of the water element, be it in a bowl of clear water or an essential oil air spray. Fresh flowers are always an excellent addition for this ritual.

2. Clearly formulate you intent, which is twofold: to clear any negative residues from the previous owners, and to imbue the house with the most positive energies you desire. A simple declaration such as "I am happy to start a new journey with you, filled with love, happiness,

and joy. I honor all the memories that you hold but would like to release everything that does not serve my intent. Thank you."

3. For this specific ritual, it is good to use your voice to express your intent, so find a way to do this in a way that feels natural and comfortable. I find that sometimes it is easier to voice things when you use sound activators, such as singing bowls or bells, for example. You can start by ringing the bell or striking your sound bowl, hovering it close to your throat in order to activate your voice (if you feel particularly shy about speaking up), then simply state your intent out loud.

4. Light the tip of your smudge stick from the candle and extinguish the flame. Take a few deep breaths and slowly smudge your energy.

5. Face the cardinal directions to express gratitude for your new home. Ask your spiritual helpers to be present during this smudging and blessings ceremony.

6. Start smudging your new home at the front door. Slowly smudge the whole house, clockwise, until you come back to where you started at the front door. Take a moment to bless your front door and energize it with your clear intent for only the most auspicious energy to enter the house. Smudge it thoroughly, both inside and out.

7. Come back to your altar and extinguish the smudge stick.

8. Take the water bowl or the essential oil mix and symbolically spray the main corners of the house. This step brings the energy of the water element as the source of renewal and abundance in your new home.

9. The last step is to take your five crystals and place them in the following manner: one crystal stays on your home altar, and the other four are placed one in each of the four cardinal directions. Have a clear intent for them to ground the energy of blessings and happiness into your new home.

10. Come back to your altar. You might want to use more sound to activate the energy, such as use the singing bowl in each corner, or just above your altar.

11. Express your heartfelt gratitude for your new home and envision it filled with laughter, joy, and sunlight. See a golden light expanding from your heart and filling the whole house with the energy of love and gratitude.

12. Take some time to sit in silence and see what insights are coming to you. You might feel that some corners of the house hold some sticky or heavy energy, so make a note of that and plan for a smudging ritual to clear this energy. You might also get insights into some specific changes the house is asking for, such as a new color for your front door. Or, maybe you will just be enjoying the delightful feeling of basking in a sweet glow of a new beginning.

Energizing your career

A smudging ritual to energize your career is easy to perform. Choose this ritual when you are looking for a new job, or when you need more inspiration and enthusiasm for your existing career. For this specific smudging, you need to smudge just two areas—your front door and your office. If you want to go deeper, you might also choose to smudge the Career feng shui area of your home, which is the North area.

Time it takes: 5 to 10 minutes

Best time of the day: Between 7 a.m. and 9 a.m.

Best time period: Last quarter of waxing moon

Best day of the week: Wednesday

Additional supplies needed: Symbol representing your ideal career, a blue candle

1. Set your smudging supplies on your altar. Add an item that represents the career you want or a specific quality you want to cultivate in your present job. This can be an image that evokes this energy for you, an item that you associate with success, or just specific words clearly written on a piece of paper.

2. Focus on your intent by clearly expressing it either in your mind or out loud. Light the candle, take a few deep breaths, light the smudge stick, and smudge yourself.

3. Go to your front door and smudge it thoroughly, both inside and out. Be sure to also smudge your main entry. If you have a busy closet by your front door, open it and carefully smudge it. Make a note to clear it, too, because busy closets in the main entry block the flow of good energy.

4. Go to your office and smudge it thoroughly, clockwise, starting from the office door. Focus on corners and lower areas closer to the floor.

5. Should you choose to smudge your feng shui career area, smudge it after you have smudged your office.

6. Come back to your altar to extinguish your smudge stick and spend some time in silence.

7. Take the special item connected to your intent from your altar and place it on your office desk.

8. Let the blue candle go out on its own.

Entering a new relationship

Consciously entering a new relationship is a powerful act. It contributes to a deeper, more satisfying experience; allows more ease in dealing with potentially difficult moments; and contributes to a more harmonious flow of energy overall. Declaring your intent for entering a new relationship is like setting a roadmap for the energy to follow—it helps you, as well as the energy of the relationship, to stay on the intended path. While declaring the intent for the relationship is ideally done by both partners—especially if you are moving in together— the smudging ritual can be performed by one person.

In this ritual, you will be smudging your whole home, then grounding the energy in three areas—the bedroom, the living room, and the love and marriage feng shui area, which is in the Southwest area of your home.

Time it takes: 20 to 30 minutes

Best time of the day: Between 11 a.m. and 1 p.m.

Best time period: Waxing moon

Best day of the week: Wednesday

Additional supplies needed: Four grounding crystals, a newly framed photo of you and your beloved, fresh pink or red flowers

1. Start at your altar where you have placed all your smudging supplies, as well as a newly framed photo of you and your partner. You can also choose to write you declaration of intent, or maybe even create a playful vision board expressing your goals and dreams for this relationship. Place on your altar four crystals or stones with grounding energy, such as

hematite, black tourmaline, black onyx, or smoky quartz. Find a beautiful way to display your fresh flowers.

2. Light the candle, take a few deep breaths, and focus on your intent. Say a prayer or a blessing, ask for help and guidance with the most auspicious path for this relationship. Light the smudge stick and slowly smudge your energy.

3. If you are smudging a home where both of you will live, start at the front door; if you are smudging your own space, start at the center, which is the heart of the home. Slowly and mindfully, smudge your whole home, proceeding clockwise. Spend a bit more time and repeat your blessings and intentions while in the bedroom, living room, and your love and marriage area.

4. Come back to your altar and extinguish the smudge stick. Let the candle follow its course, if possible. Take the photo from your altar and place it in your love and marriage area. Place your four stones in your bedroom, one in each cardinal direction, with the intent of clearly grounding the energy of love and blessings for this relationship. Leave the flowers on your altar or bring them into your living room.

The birth of a baby

The birth of a baby is a sacred occasion celebrated in all cultures. In the tradition I grew up with, it was customary to keep the baby and the mother protected from the outside influences for at least a month, so only a few people would see the baby in his or her first month. I remember feeling quite surprised to see many of my North American friends bringing their newborn babies to malls and restaurants. The energy of the new baby is wide open and receptive, so it is important to allow as gentle a transition as possible for the newborn to become accustomed to our noisy world. It is also important to create a clear, fresh energy in the baby's room to support the strength of his or her precious energy.

The intent for the birth of a baby smudging ritual is to bless both the mother and the child, to express the love and support of family and friends, and to create a space with supportive energy. This ritual can be done in the first thirty days of baby's arrival, and it can be performed by one of the parents or grandparents, or by a close friend. There are two ways to perform this specific smudging ritual—you can have it outdoors in a beautiful place in nature, if the weather allows, or you can have it in the baby's room. No matter which way you choose, note that it is good to first do a basic smudging ritual to cleanse the energy in the baby's room before the baby arrives.

You can have this ceremony in a very small circle of just a few people, or you can invite a larger community of friends. Ask your inner guidance to know what is best for you and the baby.

Prepare for this ritual by choosing one or several crystals to charge with the energy of the blessings; I suggest rose quartz, blue lace agate, or clear quartz. Any other items that will express your love for the baby and the energy of blessings are welcomed in this ceremony. You will also need two newly framed photos: one photo of the baby, and one photo of the parents. An essential oils mix of lavender, mint, and a citrus oil, be it orange, lemon, or lime, can work well to finalize the ceremony. Or, you can use a simple lavender essential oil mist. A singing bowl or a bell of good quality can help, too. You can make this ceremony as elaborate or as simple as your heart desires, but it is best not to make it go for too long.

Time it takes: 10 to 15 minutes

Best time of the day: Between 7 a.m. and 9 a.m. or 5 p.m. and 7 p.m.

Best time period: New or waxing moon

Best day of the week: Saturday

Additional supplies needed: Crystals, two newly framed photos, singing bowl or bell, lavender essential oil spray

1. Place the candle, the smudging supplies, the crystal(s) of your choice, the two photos, the singing bowl or bell, and the essential oil spray on your altar. This can be an impromptu outdoor altar, your home altar, or a new small altar in your baby's room. On or near the altar, you can also place several special gifts given to the baby by family and friends.

2. Light the candle and take a moment to connect to your intent. Ask your spirit guides, angels, or other helpers to grant their blessings for the ritual. If you decided to use a bell or a singing bowl, start by sounding it. This will announce the beginning of the ritual.

3. Light the tip of your smudge stick and then extinguish the flame. Smudge your energy while breathing deeply and staying focused on your intent. If you are in a circle of friends, step into the center of the circle to connect to the sacred directions. You can also do that from the center of the home or the center of the room. Lightly smudge the space where the ritual is being held, be it the baby's room or the outdoor place.

4. Approach the mother and the baby and ask if you can smudge them. Gently smudge the mother and the baby and be mindful to keep the smudge stick as far away from the baby as

possible. Use your hand or a smudging feather to gently fan the smoke in the direction of the baby. While smudging, say a simple blessing for the baby and the mother. You can prepare your blessing in advance, or just use the most loving words that naturally come to you in that moment.

5. After you have smudged the mother and the baby, you can ask if anyone present at the ritual wants to be smudged. Smudge those who said yes, and ask if anyone present wants to express their love and wishes for the mother and the baby.

6. Come back to your altar to extinguish the smudge stick. Take the crystals in your hand and repeat your blessings, along with a simple thank you. Ring the bell or the singing bowl again. This concludes the blessing ceremony for the new baby.

7. If possible, leave the candle to burn at its own accord. Place the two photos—the one of the baby and the photo of the parents—in the baby's room. Place the crystals that were charged during the ritual close to the photos. Use the lavender essential oil mix that was charged during the ceremony to daily clear the energy in the baby's room.

Before and after major celebrations in your home

If you like to hold frequent parties and celebrations in your home, it might be very helpful to smudge the energy before—and especially after—the party. It is important for our homes to hold strong energy that is individual to our own well-being. This energy develops in time by creating thousands of energetic connections between you and the home. The stronger these connections, the more supported you feel in your home.

When a home experiences frequent parties, especially big parties with various people, the myriad of subtle energetic connections you have with your home can weaken. In order to strengthen your home's ability to hold nourishing energy for you and your family, it is best to clear the vibrations of many people that have visited your home.

Here are two rituals that will help you take charge of the energy created before and after a big party in your home.

Smudging ritual before a major party in your home

The intent for this ritual is to create a happy and joyful energetic container for a really successful party. You are smudging to let go of any residues that might prevent a festive mood and to establish very clear boundaries. For example, you can intend to create very active energy in the spaces where the party will be held and at the same time be very clear that you do not want this energy to leak into your bedroom, or your children's bedroom.

Time it takes: 5 to 10 minutes

Best time of the day: Between 3 p.m. and 5 p.m.

Best time period: The day of the party

1. Set your smudging supplies on your altar. Light the candle. Take a deep breath and clearly formulate your intent. Hold the tip of the smudge stick over the candle. Once ignited, wait a few seconds and then extinguish the flame at the tip of your smudge stick so that it smolders.

2. Start smudging at your front door. Intend for any slow energy to leave and be replaced by energizing and vibrant energy for your celebration. You can visualize it as sparkling golden energy adorning the walls with joy.

3. Encircle with your smudging the whole area where people will be spending time. This will create an energetic cocoon that does not leak energy to the private spaces in your home. For example, from the front door you might continue to the living room, then the kitchen, the patio, the dining room, or any other area where guests will be spending time.

4. Come back to the front door to complete the energetic circle. Smudge your front door.

5. Return to your altar and extinguish the smudge stick.

6. Visualize a bright sun in your heart outpouring strong rays of light that reinforce and brighten the energetic "party adornment" you placed on your walls in Step 2. You can even intend for them to be active for a specific amount of time, let us say 5 hours. Say a short prayer welcoming your guests to your happy home with an open heart.

Smudging after a major party in your home

The intent for this ritual is to clear any energetic residues that might not be beneficial for you and your family. These energetic residues can result from obvious actions such as people arguing, or expressing sadness or anger. It can also be a more subtle energy not connected to a specific outward expression, but still not needed in your home.

Time it takes: 5 to 10 minutes

Best time of the day: Between 7 a.m. and 9 a.m.

Best time period: The morning after the party

1. Set your smudging supplies on your altar. Light the candle. Take a deep breath and clearly formulate your intent to clear any residues not in alignment with the happiness of your home. Hold the tip of the smudge stick over the candle in order to light it. Once ignited, wait a few seconds, then extinguish the flame at the tip of your smudge stick so that it smolders. Smudge yourself to clear your energy.

2. Start at the area farthest from the door where the party was being held. This can be your dining room, your kitchen, or your living room, depending on your specific floor plan. Smudge slowly and mindfully, taking deep breaths and focusing on any low, sad, or just chaotic energy to be released.

3. After you have smudged all areas that were used for the party, go to the front door and smudge it thoroughly. You can even open it a bit, if possible, and smudge around it.

4. Standing at your front door facing outward, send love to all party participants. Visualize the bright sun in your heart pushing outward and dissolving any cords or attachments that might have been formed during the party. Strengthen the energy of your front door with the intent for love and protection.

5. Return to your altar and extinguish the smudge stick. Say a short prayer thanking your helpers for keeping the energy in your home clear and bright.

Clearing after a big family argument

It is very important to clear the energy in your home after a big argument. It is obvious for us to clean the space if we have broken any object, for example, or spilled the containers of a jar. However, we pay little attention to how the house is impacted by our negative emotions. Cleaning your home energetically after the expression of heavy emotions is as important as the physical cleaning of the space.

The energy of anger, sadness, aggression, or any other heavy emotions that have been released during an argument will stay in your home and even take hold of its energy if you do not do the clearing. Happy homes are created by happy experiences. Unhappy homes are created by negative experiences left uncared for. Smudging your space after a big family argument will help you keep the energy in your home clear and happy.

This ritual is powerful if you do it by yourself. If anyone else involved in the argument wants to participate in the smudging ritual, you can say yes; just be sure the intent is on clearing the energy of the space, not on further clarifying any lingering issues.

Time it takes: 10 to 20 minutes

Best time of the day: Between 11 a.m. and 1 p.m.

Best time period: As soon as possible after the argument

Additional supplies needed: Rose quartz crystal, a calming essential oil mix of your choice (sage, jasmine, and neroli are suggested), fresh flowers, bell or singing bowl (optional)

1. Set your smudging supplies, your rose quartz crystal, your essential oil spray, and the fresh flowers on your altar. It might be very helpful to use the power of sound in this specific ceremony, such as a bell or a singing

bowl. You will also be working with the power of natural elements in this ritual. It is helpful if you do this ritual slowly and clearly visualize and connect to each element.

2. Light the candle. Take a deep breath and state your intent to clear all the negative energy from this traumatic event. Spend some time trying to release and clear your own emotions and forgive all parties involved in the argument. Journaling beforehand can help you center your energy and understand your own role in all that happened.

3. Hold the tip of the smudge stick over the candle in order to light it. Once ignited, wait a few seconds then extinguish the flame at the tip of your smudge stick so that it smolders. Smudge your energy slowly by starting at your heart. Declare out loud your intent to clear any painful energies and bring peace and clarity into your home.

4. Go to the center of your home or the room where the argument took place. Invoke the help of all seven directions, starting with the East direction and completing with your heart.

5. Begin by smudging the East area of your home or of the room where the argument took place. This is the direction of the rising sun and of welcoming new beginnings. This is also the area of Health and Family in feng shui, and the energy of a gentle young tree representing the Wood element. Connect to the freshness, gentleness, and kindness of this energy; to the hope and trust the young tree has in being supported by life. Invite this energy into your home and into your heart.

6. Move on to smudging the South area and invoke the power of Fire to burn any negativity. In feng shui, this is the direction of truth and illumination. Working with its element of Fire helps burn anything that is not in alignment with your higher truth. Invite the power of Fire to gently and thoroughly clear all negativity in your home.

7. Ask for calmness and clarity while smudging the West area. Imagine the still waters of a deep lake and the peace and contentment that this energy brings. Ask the still lake waters to calm your heart and bring wisdom and compassion for all that happened. In feng shui, this is the direction of true creativity, which cannot be sustained when our energy is in conflict. It is expressed both by the energy of the lake, as well as by the clarity, coolness, and sharpness of the Metal element.

8. Move on to the North area, which brings a strong Water element to your ritual. This is the direction of rebirth, of coming forth by being purified in deep moving waters of truth and inner courage. Ask to be washed anew, to be refreshed by clear sparkling waters of the deep and ever-moving ocean. Imagine having your energy, as well as the energy of your home, gently washed by the most clear and healing blue ocean waters. In feng shui, this direction holds the expression of your path in life, and of the responsibility to walk your path in integrity, clarity, and with an open heart.

9. Spread your arms toward the sky, honoring the Father Sky, and inviting celestial energies to shower your home with wisdom and joy.

10. Touch the ground to express your love for Mother Earth and feel her never-ending love and support; feel her motherly energy always there for you.

11. Come back to your altar and extinguish the smudge stick.

12. Place your hands over your heart. Visualize the blessings from all directions purifying the light in your heart, and then see them outpouring to fill your whole home.

13. Ring the bell or the singing bowl at your altar, as well as in each corner of the room or the whole house. Return to your altar and ring the bell again.

14. Use the essential oil spray to first mist yourself, then to mist and refresh the energy in the whole house.

15. Come back to your altar and take your crystal, holding it to your heart. Express your deep gratitude for all the help with this ceremony.

16. Let the candle go out on its own accord. You can choose to place a flower from your altar bouquet in each room. Take some time to journal any insights or sit in silence. Keep the rose quartz crystal with you for the next couple days, either in your pocket, purse, or on your bedside table.

When your relationship has ended

The end of any relationship is painful. Our energy tends to get attached to people we care for and spend a lot of time with, whether in the form of a romantic relationship, a close friendship, or even a longer working relationship that was fruitful but not possible to continue any longer. Letting go of these cords can bring an array of painful emotions that need to be felt in order to be cleared.

A smudging ritual with the intent of clearing attachment cords can be very helpful in this process. It might be necessary to do this ritual several times because of the strong emotions it can bring. It is also helpful to do a lot of journaling both before and after this smudging ritual. Allow yourself an open window of time after this ritual to fully feel all emotions and memories that are asking to be felt and released. If you rush into any activity shortly after this ritual, its power will be diminished considerably.

Please also note that depending on where you are in the process of grieving, your energy might not be ready to let go yet. Trust your inner guidance and see if the timing is right for you to do this ritual. If you need more time before fully letting go, you can encourage your heart's healing by performing a daily smudging of your personal energy and your bedroom. Apply this ritual only when you know you want to move on.

There are two specific items you will need for this ritual, so spend some time deciding on what to choose. The first item is one that has a direct connection to the relationship you are letting go of, and the second item represents its energy. Here is the difference between the two: A photo of you and the partner you are separating from is an item that has direct connection to your relationship. A photo of a couple enjoying swimming in the ocean or a big heart-shaped rose quartz crystal are examples of items that represent the energy of a love relationship. Our goal with this ritual is to clearly separate the two in order to gently speed up the healing of your heart. This will help your heart stay open to receive the same energy from the Universe, just with other people.

Time it takes: 15 to 30 minutes

Best time of the day: Between 3 p.m. and 5 p.m.

Best time period: Full or waning moon

Best day of the week: Thursday

Additional supplies needed: An item directly connected to this relationship, an item representing the energy of desired relationship, a crystal of your choice

1. Set the smudging supplies on your altar. Place your candle in the center and the crystal of your choice near it. Rose quartz, rhodochrosite, or a pink tourmaline are ideal. The item with the direct connection to your relationship is placed to the left of the candle. Place the item that represents the essence of this relationship to the right of the candle.

2. Light the candle and take a few deep breaths. Ground your energy, open your heart, and allow your emotions to flow. Ask God, the Great Spirit, the Divine Mother, or any other deities you are working with to help you release these cords fully. Ask for help in releasing

these cords with love. Declare your intent to stay open to the energy of love by releasing the sadness or pain connected to the end of this relationship.

3. Light the smudge stick and smudge yourself thoroughly.

4. Smudge your whole space, clockwise, starting in your bedroom. The bedroom is the space that tends to hold a lot of our unexpressed emotions, so clearing it thoroughly is very important. If you have spent a lot of time with this person in your home, pay special attention to smudging the areas where you spent most of your time together. Let the energy flow, feel whatever needs to be felt, and do not control your emotions.

5. If you need to stop, you can place your smudge stick in its bowl on your altar and go through whatever your body is asking for in order to release the energy of pain. You can feel a lot of anger that takes you by surprise, or a deep pit of sadness in your solar plexus area. You might feel like sobbing and mourning all the hopes you had for this relationship. Let this energy flow. All you need to do is feel it, be its friend, and be there for this pain. The wisdom and release of energy is not so much in its specific expression, even though that might be very cathartic, too. The release happens by deeply feeling the emotion in your body without the need to do anything about it—no fixing, no suppressing, not even helping it move. Your body knows when and how to move it; all you need to do is be there for it. If it feels right for you to stop the ritual and go for a walk to process and feel all that is coming up for you, please do that.

6. When the time is right, continue the ritual. Smudge your whole space clockwise and come back to your bedroom, the place where you started. Smudge yourself again, reaffirming that you are being cleansed of pain.

7. Go to your altar and extinguish the smudge stick.

8. Ground your energy by sending a column of light from your sacral area deep into the womb of Mother Earth. Send your gratitude for all that you are gifted with in this earthly life. Then send a column of light energy from your heart all the way to the central sun. You can visualize this energy high above your head, showering you with wisdom and peace. Next, focus on your heart and blend these two sacred energies—the nourishing, caring, and loving energy of Mother Earth with the peaceful and clear guidance from the Father Sky.

9. Affirm your desire for love and fulfilling connections in your life. Ask for help in releasing all lingering sadness. Know that you are being heard and trust that what you want is coming to you. You are a precious, loved, and unique child of the Universe; all you need to do is clearly ask. You do not need to know how this will happen, you only need to trust.

10. Take the item that has the direct link to the relationship you want to let go of and find the best, most honorable way to discard it.

11. Place the item that expresses the energy of what you want as a focal point on your altar. Place the crystal near it. You can let the candle burn for a bit longer, or until it completes its natural course.

Smudging after someone's death

Dealing with the energy of death still feels like taboo in most Western countries. We hide away most things related to death and prefer not to think about it. We drown the terrifying fear of death in a myriad of obsessions and distractions. Not many of us are truly connected to the mysteries of the Universe, the mysteries that give us the needed certainty that life does not end. Viewing this world as the only world in existence, we are as afraid to live as we are afraid to die.

Dealing with the death of a friend, a family member, a neighbor, or even someone you hardly know will bring powerful emotions to the surface. The sudden loss of life can shake us to the core. Sacred rituals were developed in many ancient cultures to help bring a sense of order to these powerful emotions, as well as to help cope with the grief that inevitably follows. These cultures also used rituals to clear the energy of the space if someone died at home.

You can use this smudging and space-clearing ritual when you or someone close to you is affected by someone's passing. This ritual can take place in the home of the one who has passed away, or in the home of a friend who is grieving. Performing this ritual in a close circle of people is always more powerful. At times, though, you might need to do it by yourself. Ask your inner guidance and sense what feels right for your specific situation.

As grief comes in many stages, this ritual can be used, with some modifications, as often as needed. Before this ritual, it is good to smudge the space thoroughly, especially if this ceremony is done in the home of the one who has passed away.

Time it takes: 30 minutes to 1 hour

Best time of the day: Between 6 p.m. and 9 p.m.

Best time period: Waning moon or when necessary

Best day of the week: Sunday or when necessary

Additional supplies needed: A photo of the person who passed away, fresh flowers, bell or singing bowl, an essential oil mix (clary sage, sandalwood, frankincense, ylang ylang, jasmine, lavender, grapefruit, or orange oil are suggested)

1. Create an altar on which to gather all your smudging supplies. Place the picture of the deceased person in the center of the altar and the vase with fresh flowers close to it.

2. Light the candle. Take a few deep breaths and bring to mind the energy of the person you are grieving. See them as happy and content, see them in light energy. Recall your times together and try to bring to mind any joyful moments you experienced. Say a prayer for both of you. Ask the Spirit to bless and take care of the Soul of the person who has moved on. Ask to be granted strength and heart healing to help you continue on your own journey.

3. Light your smudge stick and smudge yourself, slowly and mindfully. Start at your heart. Visualize it opening gently, and express its voice and its feelings. Whatever comes up, let it be, while gently and consistently smudging your energy. The sacred herbs help release energetic layers of grief that accumulate, especially around one's heart and solar plexus area. It is also important to thoroughly smudge two more areas of your body—the area above your head and the soles of your feet. Smudging above your head helps open you up to higher wisdom and releases any negative thought forms that might be hovering over you. Smudging the soles of the feet is very important in order to allow a stronger connection to Mother Earth and her healing, grounding energies.

4. If you are doing this ceremony with others, offer to take them through the same smudging of their personal energy. This can take some time, as everyone is different in expressing their feelings and emotions. Keep breathing, keep smudging slowly and mindfully, and ask the Spirit to help bring the necessary healing for everyone in this moment.

5. Smudge the photo of the deceased person, and take flowers from the vase and place them around the photo.

6. Go to the center of the space, or the center of the circle, and invoke the blessings of all seven directions. Start by facing East and then gently wave the sacred smoke in each direction. Invoke these blessings for everyone—for the person whose soul moved on in its journey, for you, for the people present at the ceremony, and for everyone in this world who is grieving.

7. Return to the altar and place your smudge stick in its bowl, extinguishing it. Mist the essential oils into the air. You can also mist your own energy, as well as the energy of all present.

8. Pick up the bell or singing bowl and ring it gently. Invite the energies of life and death to co-create a new path for all those present at the ceremony, a path walked in love, kindness, and purpose.

9. Sit in silence and allow insights to come. Be grateful for all the unseen help you received during this ceremony.

Asking for clarity

There are times in everyone's life when things feel unclear or confusing. Many techniques can be used to bring clarity back into one's life, such as asking yourself hard questions, meditating, spending time in nature, praying, and clarifying your intent. A smudging ritual can help clear the energy of confusion and chaos, as well bring a sense of peace and calm.

A celestite crystal can be helpful for this ceremony, as well as a small bowl of pure water with a few drops of lavender, lemon, and mint essential oils in it. The only areas to be smudged in this ceremony are the bedroom, the front door, and the main entry area. You can decide to smudge the whole house, of course, but these specific areas need to be smudged thoroughly and slowly.

Time it takes: 15 to 20 minutes

Best time of the day: Between 5 a.m. and 7 a.m. or 11 p.m. and 1 a.m.

Best time period: New moon or whenever needed

Best day of the week: Tuesday

Additional supplies needed: Sheets of paper to journal on, pen or pencil, a celestite or a clear quartz crystal, a bowl of water with a few drops of essential oils

1. Before this smudging ritual, take several sheets of white, unlined paper and express in writing all you feel about your current state. What brings you the most confusion? What is the ideal solution to this—how would you like to feel? Other than clarity, what do you want to ask for help with? You can do automatic writing, where you quickly write down everything that first comes to mind without consciously processing it; you can draw; or you can just write any words without connecting them in a specific order. See what the moment is asking you to put in writing. Once finished, place the papers and writing instrument on your altar.

2. Place the celestite crystal on top of your papers. Sprinkle yourself and the altar with the essential oils water from the bowl.

3. Light the candle and take a few deep breaths. Ask your Spirit guides, helpers, and angels to be with you. Ask for clarity.

4. Light the tip of the smudge stick from the candle and blow out the flame. Mindfully smudge yourself.

5. Go to the center of your home and invoke the help of all seven directions—East, South, West, North, Sky, Earth, and Heart.

6. Start by smudging your bedroom. Do the smudging ritual very slowly, paying attention to all bedroom corners, windows, as well as the area under the bed. Open your closet and smudge it thoroughly and carefully, keeping a reasonable distance from all the items in your closet. Remember to use a bowl under your smudge stick so as not to drop any ashes. Be open to receive any insights, as your bedroom holds a lot of your unconscious energy and it can reveal things that you were unaware of before. Clearing and shifting the energy in your bedroom is the first step in order to bring clarity into your life. Do you store things under the bed? Is your closet full and unsightly? Do you love being in your bedroom? Tune in to how your bedroom feels and where the energy feels stuck. Make mental notes as to what you can change for the better. Keep smudging slowly, from the lowest areas close to the floor to the highest you can reach near the ceiling.

7. When the process of smudging your bedroom feels complete, go to your front door. The front door is how your house absorbs vital energy needed to maintain good energy in your whole space. The quality of energy at your front door determines the quality of energy in your house. Usually, when we go through a lot of confusion, there is a lot that can be cleared in the main entry and near the front door in order to allow fresh energy to come in. Open the front door and smudge it inside and out.

8. Open the main entry closets and smudge them. Feel if this area is blocked and sense what changes you can make to shift the energy. Mindfully smudge all corners of your main entry, starting at the lower level and proceeding to higher levels closer to the ceiling.

9. When you feel complete, return to the center of your home. Take a moment to ground your energy and express gratitude for this process.

10. Go back to your altar and extinguish the smudge stick. Place the celestite crystal in the water bowl and sprinkle your energy with the essential oils water.

11. Take the water bowl and go back to your bedroom. Use this water to sprinkle fresh energy around your bedroom. Do the same process with your front door and the main entry.

12. Return to your altar and place the bowl back. Take the crystal out of the water and place it near the candle. Let the candle burn its natural course. Once the flame is out, take the crystal and carry it with you for a while, either in your pocket, in your purse, or by your bed stand. Make a note of the changes you can make in your bedroom and front door area and plan to make these changes as soon as possible.

13. Dip the papers you wrote on in the bowl of water, and wait until they start to dissolve. Compost them.

Before big changes

The attitude toward change fully reveals our complex human nature. We long for change and we are terrified of it at the same time. We want change in some areas but not in others. We want to feel in control of any changes and manage them according to our expectations. Of course, when life brings change, it is mostly out of our control. The best we can do is stay calm, clear, and heart-centered, and keep our energy open and connected to Mother Earth's energy.

This simple smudging ritual will help bring some peace, trust, and centeredness to your energy. It can be performed daily to help clear any fear residues and soothe your heart with the power of natural elements.

Time it takes: 5 to 10 minutes

Best time of the day: Between 7 a.m. and 9 a.m.

Best time period: Anytime

Best day of the week: Any day

Additional supplies needed: A symbol representing the energy of hope and items to represent the five elements (Wood, Fire, Earth, Metal, and Water)

1. Place a symbol on your altar that represents for you the energy of hope and trust. It can be a family photo or any other inspiring visual such as an image from nature. It can also be a specific crystal, or even a story or a book that touched your heart.

2. Surround it with items that represent the five feng shui elements. Water can be represented by a small bowl of water, Earth by crystals, Wood by a small vibrant plant or a bouquet of flowers, Metal by a couple of new coins, and Fire by the candle. Find a way to display them so that it brings peace to your heart and expresses your creativity. You want your nature elements mandala around your symbol to have a joyful and colorful energy. Do it mindfully, and do not rush.

3. Light the candle, then take a few deep breaths. Declare your intent to welcome the coming change with strength, calm, and an open heart. Light the tip of the smudge stick from the candle flame and blow it out so that it smolders. Smudge your energy slowly and mindfully.

4. Go to the center of your home, also known as the heart of the home. Ask for help from all seven directions and gently smudge in each direction by standing in the center of the house.

5. Next, go to your closet, open it, smudge it lightly and carefully, and then take one or several items from it that you have not used in a long time. Put them close to the door so you can let them go as soon as possible.

6. Come back to your altar, smudge it, and extinguish the smudge stick. Let the candle burn while you journal about your experience. Leave the mandala on your altar and repeat this

smudging ritual as often as you need to. Be sure the items representing the elements of Water and Wood are kept fresh on your altar. This might mean you change the water in the bowl every morning, as well as bring fresh flowers as needed.

7. Give the unused items from your closet either to your neighbor or to a local thrift store. Do not delay until you have a big pile of clothing or other items to give; just give them away as soon as possible. This will promote the release of stagnant or blocked energy from your personal energy field and help open you to change.

Opening to love

If you find yourself in a time when your heart is longing for more love, this ritual will help open the channels for love energy to flow into your life. I firmly believe that love tends to flow toward itself, meaning it goes to where there is already love dwelling and expressing itself. I also believe that love does not really need to be "attracted" into your life; it just needs to be opened up to and allowed to flow freely. Love does not come from far away. It is always there, so it is all a matter of opening yourself to love and allowing it in.

This ritual brings best results when you clearly hold your intent both on magnifying the energy of self-love and on attracting your partner. One has to be there in order to attract the other. Express your intent for this ritual in the way that feels most meaningful to you. Make it as simple or as elaborate as your heart is longing for. However, do not forget to focus on both asking for the ability to love your heart more deeply and to attract your beloved.

In this ritual, we will be smudging two areas—your bedroom and your love and marriage feng shui area, which is the Southwest area of your house.

Time it takes: 15 to 20 minutes

Best time of the day: Between 3 p.m. and 5 p.m.

Best time period: New moon

Best day of the week: Saturday

Additional supplies needed: Two rose quartz crystals, a personal amulet such as a necklace or pendant, and a visual of the love energy you want to attract

1. Set all items on your smudging altar. Use a pink candle, if you can. Light the candle, then take a few deep breaths. Express out loud your intent for this ritual.

2. Light the smudge stick as usual and smudge your personal energy. Focus on your heart area to sense if the energy there is open or closed. Feel whatever you are feeling in your heart in this moment—be it fear, longing, or sadness—and allow yourself to feel it deeply. Breathe, and do not rush. Strong emotions might come, so stay with them and do not feel the need to change them.

3. Go into your bedroom and slowly smudge it, clockwise, starting at the door. Pay special attention to the area around the bed—smudge both above the bed and under it. This is the time to have your bowl under the smudge stick, as you do not want any ashes falling on your bed. If you have a busy closet in the bedroom, open it and smudge carefully. Also make a note to clear the closet of any clutter. Inviting love into your life means making space for it, and, feng shui-wise, you literally have to make space for it in your home, especially in your closets.

4. Once the bedroom is smudged, go to the love and marriage feng shui area and smudge it thoroughly. Pay attention to how this area feels and what it might be asking for. Is there an inviting quality of energy, loving and open, or does this area feel closed and unfriendly? Feng shui-wise, this area expresses, as well as nourishes, the energy of your love relationship. While smudging it, see if you receive any insights on how to create a more loving, happy energy there.

5. Come back to your altar and extinguish the smudge stick. Pick up your two rose quartz crystals, hold them in your hand, and energize them with your intent for love. Take your visual from the altar, and along with the crystals, place them in your love and marriage area. If it feels better to you, place your visual in your bedroom.

6. Wear your personal amulet often and keep its energy fresh by lightly smudging it and placing it overnight on your altar. Trust that love is already there for you, patiently waiting for a full "Yes!" coming from a deeply surrendered and open heart.

ABOUT THE AUTHOR

Rodika Tchi is a professional feng shui consultant, teacher, and writer. She has been consulting for both residential and commercial property owners for over 18 years, and has helped numerous clients create harmonious homes, as well as quickly sell houses or improve businesses.

The creator of KnowFengShui.com and the long-time feng shui expert for The Spruce, Rodika has been interviewed by *ELLE Decor*, *Style at Home*, *Canadian Living*, *The Globe and Mail*, and many others. Find more about Rodika on TchiConsulting.com.